John Lubbock

On British Wild Flowers Considered in Relation to Insects

John Lubbock

On British Wild Flowers Considered in Relation to Insects

ISBN/EAN: 9783337107475

Printed in Europe, USA, Canada, Australia, Japan

Cover: Foto ©berggeist007 / pixelio.de

More available books at **www.hansebooks.com**

NATURE SERIES

ON

BRITISH WILD FLOWERS

CONSIDERED IN

RELATION TO INSECTS

BY

SIR JOHN LUBBOCK, BART., M.P., F.R.S., D.C.L., LL.D.

PRINCIPAL OF THE WORKING MEN'S COLLEGE; PRESIDENT OF THE LONDON CHAMBER OF
COMMERCE; AND VICE-CHAIRMAN OF THE LONDON COUNTY COUNCIL

WITH NUMEROUS ILLUSTRATIONS

London
MACMILLAN AND CO.
AND NEW YORK
1890

RICHARD CLAY AND SONS, LIMITED,
LONDON AND BUNGAY.

Printed 1874
Reprinted 1875, 1882, 1885
New Edition 1890.

PREFACE.

IT is not without much diffidence that I venture on
the present publication. For though as an entomo-
logist I have necessarily been long familiar with our
common wild plants, I had made no serious study of
Botany until recent researches brought prominently
before us the intimate relations which exist between
flowers and insects. My observations and notes on
this subject were originally prepared with the view
of encouraging in my children that love of natural
history from which I myself have derived so much
happiness, but it was suggested to me that a little
book such as the present might perhaps be of use
to others also.

Sprengel, in his admirable work, "Das entdeckte
Geheimniss der Natur," published as long ago as the
year 1793, was the first to show how much plants are
dependent on the visits of insects, and to point out
that the forms and colours of flowers are adapted to
ensure, and profit by, those visits. His work, however, did not attract the attention which it deserved,
and our knowledge of the subject made little pro-
gress until the publication of Mr. Darwin's researches,
to which I shall continually have occasion to refer.
Dr. Hermann Müller in his "Die Befruchtung der
Blumen durch Insekten," has brought together
the observations of previous writers, and added

to them an immense number of his own. Many other naturalists—for instance, Axell, Bennett, Delpino, Hildebrand, Hooker, F. Müller, and Ogle, have also published valuable memoirs on the subject, to which I shall frequently have occasion to refer; but to the works of Sprengel, Darwin, and Dr. H. Müller I am indebted in almost every page, and in spite of constant references, it is impossible for me adequately to acknowledge my obligations to them. In the systematic portion, I have followed Mr. Bentham's excellent "Handbook of the British Flora."

As far as possible, I have avoided the use of technical terms, but some were unavoidable; references for these will be found in the Index, and I have also given a Glossary of the technical terms most frequently employed.

I have to thank various friends who have been good enough to assist me, but especially Dr. (now Sir Joseph) Hooker and Mr. Busk, who have been so very kind as to look through my proofs.

In conclusion, I must add that the subject is comparatively new, and many of the observations have not yet stood that ordeal of repetition which they will no doubt experience. While, therefore, I believe that the facts will be found to be in the main correct, the inferences drawn from them must, in many cases, be regarded rather as suggestions than as well established theories. The whole subject is one which is most interesting in itself, and will richly repay those who devote themselves to it.

High Elms, Down, Kent,
September, 1874.

CONTENTS.

CHAPTER IV.

CHAPTER V.

CHAPTER VI.

CHAPTER VII.

LIST OF ILLUSTRATIONS.

. In all the figures of flowers, unless otherwise specified, the letters refer to the same parts, viz. :—pistil, *p*—style, *p'*—stigma, *st*—stamen, *f*—filament, *f'*—anther, *a*—petals, *pe*—corolla, *co*—sepals, *s.*—calyx, *ca*—ovary, *o*—honeygland, *h*—pollen, *po*.

GLOSSARY.

Anemophilous (p. 9) plants are those in which the pollen is carried to the stigma by the wind.

Anther, that portion of the stamen which contains the pollen.

Calyx (p. 27), the outer whorl of the flower.

Cleistogamous species (p. 37), are those which, besides the usual conspicuous flowers, have others which are smaller, and generally uncoloured.

Corolla (p. 27), the second whorl of the flower. In most cases this is the coloured part.

Dichogamous species (p. 28) are those in which the stamens and pistil do not mature simultaneously.

Diclinous plants (p. 28), are those in which all the flowers are either male or female, that is to say, either contain stamens but no pistil, or pistil but no stamens.

Dimorphous species (p. 29) are those in which there are two forms of flowers, differing in the relative position or length of the anthers and stigma.

Diœcious species (p. 28) are those in which the stamens and pistils are situated not only in distinct flowers, but also on separate plants.

Entomophilous plants (p. 9) are those in which the pollen is carried to the stigma by insects

Epigynous, situated upon the ovary.

Filament, the stalk of the anther.

Heterogamous plants are those which have male, female, and hermaphrodite flowers, or any two of them united in one head.

Heteromorphous species are those in which there is more than one form of flower.

Hypogynous, situated under the ovary.

Monœcious species (p. 28) are those in which the stamens and pistils are in separate flowers, but on the same plant.

Monomorphous species are those in which all the flowers resemble one another in the relative position of the stamens and pistil.

Nectary, that part of the flower which secretes honey.

Perigynous, situated around the ovary.

Petals, the leaves of the corolla.

Pistil, the central organ of the flower. It generally consists of one or more ovaries and stigmas ; the stigma is often raised on a stalk, called a " style."

Polygamous species are those which have male, female, and hermaphrodite flowers on the same or on distinct plants.

Proterandrous plants (p. 28) are those in which the stamens come to maturity before the pistil.

Proterogynous plants (p. 28) are those in which the pistil comes to maturity before the stamens.

Sepals (p. 27) the leaves of the calyx.

Stamens (p. 27) the parts of a flower which generally stand next the corolla, on the inner side. They usually consist of a stalk or filament, and an "anther" containing the pollen.

Stigma (p. 27), that portion of the pistil in which pollen must be deposited in order to fertilise the flowers.

Style, the stalk of the stigma.

Trimorphous species are those in which there are three forms of flowers, differing in the relative position or length of the anthers and stigma.

GERANIUM SYLVATICUM.

ON BRITISH WILD FLOWERS

CONSIDERED IN

RELATION TO INSECTS.

CHAPTER I.—INTRODUCTION.

THE flowers of our gardens differ much in size and colour from those of the same species growing wild in their native woods and fields: this is due partly to cultivation, but still more to the careful selection of seeds or cuttings from those plants, the flowers of which show any superiority over the others in size or colour.

Even amongst wild flowers, however, recent researches have proved that the forms and colours have been modified in a similar manner: the observations of botanists, especially of Sprengel and Darwin, have shown that the forms and colours of wild flowers are mainly owing to the unconscious selection exercised by insects, although no doubt the existence of a certain amount of colouring matter is, as we see in the autumn tints, in various .fungi, seaweeds, &c., due to other causes.

Sprengel appears to have been the first who perceived the intimate relations which exist between plants and insects; and *Geranium sylvaticum* (see p. 1) will always have an interest as being the flower which first led him to his researches. In the year 1787 he observed that in the corolla of this species there are a number of delicate hairs; and, convinced, as he says, that "the wise Author of Nature would not have created even a hair in vain," he endeavoured to ascertain the use of these hairs, and satisfied himself that they served to protect the honey from rain.

His attention having thus been drawn to the subject, he examined numerous other flowers with great care, and was surprised to find how many points in reference to them could be explained by their relations to insects.

The visits of insects are of great importance to plants in transferring the pollen from the stamens to the pistil. In many plants the stamens and pistil are situated in separate flowers: and even in those cases where they are contained in the same flower, self-fertilisation is often rendered difficult, or impossible; sometimes by the relative position of the stamens and pistil, sometimes by their not coming to maturity at

the same time.   Under these circumstances the transference of the pollen from the stamens to the pistil
is effected in various ways.   In some species the
pollen is carried by the action of the wind; in some
few cases, by birds; but in the majority, this important object is secured by the visits of insects,
and the whole organisation of such flowers is adapted
to this purpose.

To the honey are due the visits of insects; the
sweet scent and bright colours of the flowers attract
them ; the lines and circles on the corolla guide them
to the right spot; and, as we shall see, there are a
number of curious contrivances all tending to the same
object. But while Sprengel's deep religious feeling thus
gave him the clue which has thrown so much light on
the origin and structure of flowers, the comparatively
low conception of creative power which was in his
time, and, indeed, until recently, prevalent, led him to
assume that each flower was created as we now see it,
and prevented him from perceiving the real significance of the facts which he had discovered ; while
the true explanation could scarcely have escaped
him if he had possessed that higher view of creation
which we owe to Mr. Darwin.   Though he observed
that in many species the stamens and pistil are not
mature simultaneously, and that such plants therefore cannot fertilise themselves, but are generally
dependent on the visits of insects, he appears to have
considered that these visits were arranged mainly in
order to overcome the difficulty of fertilisation thus
resulting ; and hence, perhaps, the oblivion into which
his work, though so interesting and suggestive in

itself, so full of curious and careful observations, was allowed to fall. For there is an obvious inconsistency in the coexistence of two elaborate sets of arrangements, one tending to preclude, the other to effect, self-fertilisation ; in supposing that in the first place the stamens and pistil were so arranged that the pollen of the one might not fertilise the other ; and, secondly, that elaborate contrivances were devised to promote the visits of insects, and compel them to transfer the pollen from the stamens to the pistil: a result which might have been obtained so much more simply by a slight alteration of the flower itself.

It is the more remarkable that this did not strike Sprengel, because he expressly observes in one passage that, "Die Natur nicht will dass irgend einer Zwitterblume durch ihren eigenen Staub befruchtet werden solle" (Nature does not choose that any complete flower should be fertilised by its own pollen). Yet though thus so near the truth, he failed to perceive the true importance of the visits of insects. Subsequent observers, though in some cases recognising the advantage of fertilising one flower by pollen from another, did not connect these observations with Sprengel's discoveries ; and our illustrious countryman Mr. Darwin was the first to bring into prominence the fact that the importance of insects to flowers consisted in their transferring the pollen—not merely from the stamens to the pistil, but from the stamens of one plant to the pistil of another.

While then from time immemorial we have known that flowers are of great importance to insects, it is only comparatively of late that we have realised how important, indeed how necessary, insects are to flowers. For it is not too much to say, that if, on the one hand, flowers are in many cases necessary to the existence of insects ; insects, on the other hand, are still more indispensable to the very existence of flowers :—that, if insects have been in many cases modified and adapted with a view to obtain honey and pollen from flowers, flowers in their turn owe their scent and colour, their honey, and even their distinctive forms to the action of insects. There has thus been an interaction of insects upon flowers, and of flowers upon insects, resulting in the gradual modification of both.

If it be objected that I am assuming the existence of these gradual modifications, I must reply that it is not here my purpose to discuss the doctrine of Natural Selection. I may, however, remind the reader that Mr. Darwin's theory is based on the following considerations :—1. That no two animals or plants in nature are identical in all respects. 2. That the offspring tend to inherit the peculiarities of their parents. 3. That of those which come into existence, only a certain number reach maturity. 4. That those which are, on the whole, best adapted to the circumstances in which they are placed, are most likely to leave descendants.

Now, applying these considerations to flowers ; if it be an advantage to them that they should be visited by insects (and that this is so will presently be shown), then it is obvious that those flowers which, either by their larger size, or brighter colour, or sweeter scent, or greater richness in honey, are most attractive to insects, will, *cæteris paribus*, have an advantage in the struggle for existence, and be most likely to perpetuate their race.

Every garden indeed is a sufficient proof that in size and colour, flowers are susceptible of great modifications; and insects unconsciously produce changes similar to those which man effects by design.

Insects are useful to plants in various ways. Thus, a species of acacia mentioned by Mr. Belt,[1] if unprotected, is apt to be stripped of its leaves by a leaf-cutting ant, which uses the leaves, not directly for food, but, according to Mr. Belt, to grow mushrooms on. The acacia, however, bears hollow thorns, and each leaflet produces honey in a crater-formed gland at the base, and a small, sweet, pear-shaped body at the tip. In consequence, it is inhabited by myriads of a small ant, *Pseudomyrma bicolor*, which nests in the hollow thorns, and thus finds meat, drink, and lodging all provided for it. These ants are continually roaming over the plant, and constitute a most efficient body-guard, not only driving off the leaf-cutting ants, but even in Mr. Belt's opinion, rendering the leaves less liable to be eaten by herbivorous mammalia.

The principal service, however, which insects perform for plants is that of transferring the pollen from one flower to another.

I will not now enter on the large question why this cross-fertilisation should be an advantage; but that

---

[1] F. Müller has observed similar facts in Sta. Catharina. (*Nature*, vol. x. p. 102.)

it is so has been clearly proved. Kolreuter speaks
with astonishment of the *"statura portentosa"* of some
plants thus raised by him ; indeed, says Mr. Darwin
("Animals and Plants under Domestication," ch. xvii.),
"all experimenters have been struck with the won-
derful vigour, height, size, tenacity of life, precocity,
and hardiness of their hybrid productions."   Mr.
Darwin himself, however, was, I believe, the first to
show that if a flower be fertilised by pollen from a
different plant, the seedlings so produced are much
stronger than if the plant be fertilised by its own
pollen.   I have had the advantage of seeing several
of these experiments, and the difference is certainly
most striking.   For instance, six crossed and six self-
fertilised seeds of *Ipomœa purpurea* were grown in
pairs on opposite sides of the same pots ; the former
reached a height of 7 ft., while the others were on
an average only 5 ft. 4 in.   The first also flowered
more profusely.   It is moreover remarkable that in
many cases plants are themselves more fertile if sup-
plied with pollen from a different flower, a different
variety, or even, as it would appear in some instances
(in the passion flower, for instance), from a different
species.   Nay, in some cases pollen has no effect
whatever unless transferred to a different flower.
Fritz Müller has recorded some species in which
pollen, if placed on the stigma of the same flower,
has not only no more effect than so much inorganic
dust ; but, which is perhaps even more extraordinary,
in others, he states that the pollen placed on the
stigma of its own flower acted on it like a poison.
This he noticed in several species: the flower faded and

fell off; the pollen-grains themselves, and the stigma in contact with them, shrivelled up, turned brown, and decayed; while other flowers on the same branch, which were not so treated, retained their freshness.

The transference of the pollen from one flower to another is, as I have already mentioned, effected principally either by the wind or by insects. In the former case the flower is rarely conspicuous; indeed Mr. Darwin finds it "an invariable rule that when a flower is fertilised by the wind it never has a gaily-coloured corolla." Conifers, grasses, birches, poplars, &c., belong to this category.

In such plants a much larger quantity of pollen is required than where fertilisation is effected by insects. Everyone has observed the showers of yellow pollen produced by the Scotch fir. It is an advantage to these plants to flower before the leaves are out, because the latter would greatly interfere with the access of the pollen to the female flower. Hence such plants, as a rule, flower early in the spring. Again, in such flowers the filaments of the stamens are generally long, and the pollen is less adherent, so that it can easily be detached by the wind, which would manifestly be a disadvantage in the case of those flowers which are fertilised by insects. On the other hand, it is an advantage to most seeds to be somewhat tightly attached, because they are then only removed by a high wind which is capable of carrying them some distance. I say "to most" because this does not apply to such seeds as those of the dandelion, which are specially adapted to be carried by the wind.

Again, as Mr. Darwin has pointed out, irregular flowers appear to be almost always fertilised by insects.

Wind-fertilised flowers, moreover, generally have the stigma more or less branched or hairy, which evidently tends to increase its chance of catching the pollen.

Figs. 1 to 6, taken from Axell's work, illustrate

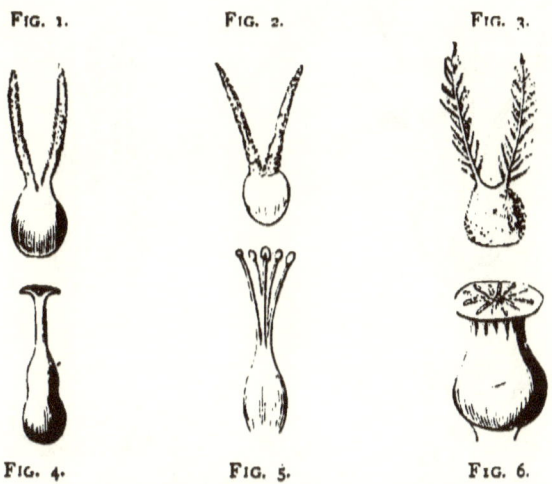

FIG. 1.   FIG. 2.   FIG. 3.

FIG. 4.   FIG. 5.   FIG. 6.

FIG. 1. Stigma of the Alder. FIG. 2.—Of the Hop. FIG. 3.—Of the Wheat; which are anemophilous. FIG. 4.—Of the Willow. FIG. 5.—Of the Flax. FIG. 6.—Of Nuphar; which are entomophilous.

this difference. In the alder (Fig. 1), the hop (Fig. 2), and wheat (Fig. 3), the pollen is wind-borne, whence they have been termed by Delpino " *anemophilous;* " while in the willow (Fig. 4), the flax (Fig. 5), and nuphar, (the yellow water lily) (Fig. 6), it is carried by insects, whence such plants have been termed " *entomophilous.*"

Even in nearly allied plants this difference is well marked, in illustration of which Axell gives the following figures taken from Maout and Decaisne's "Traité générale de Botanique":—Fig. 7 represents a section of a flower of *Plantago major*, which is wind-fertilised ; Fig. 8 of an allied species, *Plumbago*

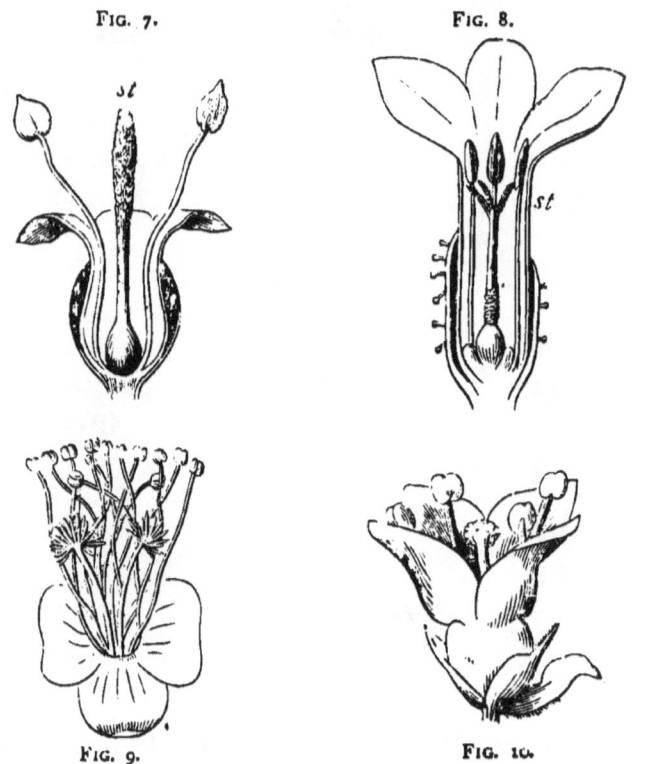

FIG. 7.   FIG. 8.

FIG. 9.   FIG. 10.

FIG. 7—Section of Plantago Major.   FIG. 8.—Of Plumbago Europea.
FIG. 9.—Flower of Poterium sanguisorba.   FIG. 10.—Of Sanguisorba officinalis

Europea, which is insect-fertilised.   Again, Fig. 9 represents a section of *Poterium sanguisorba*, which is wind-fertilised ; Fig. 10 of the nearly allied *Sanguisorba officinalis*, which is fertilised by insects.

It is an almost invariable rule that wind-fertilised flowers are inconspicuous ; but the reverse does not hold good, and there are many flowers which, though habitually visited by insects, are not brightly coloured. In some cases, flowers make up by their numbers for the want of individual conspicuousness. In others, the insects are attracted by scent ; indeed, as has already been mentioned, not only the colour [1] of flowers, but the scent also, has no doubt been greatly developed through natural selection, as an attraction to insects. But though bright colours and strong odours are sufficient to attract the attention of insects, something more is required. Flowers, however sweet-smelling or beautiful, would not be visited by insects unless they had some inducements more substantial to offer. These advantages are the pollen and the honey ; although it has been suggested that some flowers beguile insects by holding out the expectation of honey which does not really exist, just as some animals repel their enemies by resembling other species which are either dangerous or disagreeable.

Night flowers are generally white or pale yellow, these being the tints which render them most conspicuous in the dusk of evening. Thus *Lychnis diurna*, which opens by day, is red ; while *L. vespertina*, which opens in the evening, is white.

It will scarcely, I think, be doubted by any one

---

[1] In confirmation of this it is stated that when insects are excluded, the blossoms last longer than is otherwise the case ; that when flowers are once fertilised, the corolla soon drops off, its function being performed.

that *scent* is an advantage to flowers by attracting insects. No wind-fertilised flowers are scented. On the other hand, while colour is as useful as scent by day, at night it is of course less easily perceived. Hence night flowers are specially odoriferous, and there are some—such as *Hesperis matronalis, Silene nutans*, &c.—which are very sweet in the evening and yet emit little or no odour by day.

The *honey* is secreted, sometimes by one part of the flower, sometimes by another; and great variations may be found in this respect even within the limits of a single order. Thus in the Ranunculaceæ the honey glands are situated on the calyx, in certain Pæonies; on the petals, in buttercups and hellebore; on the stamens, according to Müller, in Pulsatilla; and on the ovary, in Caltha.

The real use of the honey in flowers, indeed, now seems so obvious that it is remarkable to see the various theories which were entertained on the subject. Patrick Blair thought it absorbed the pollen, and thus fertilised the ovary. Linnæus confessed his inability to solve the question. Other botanists considered it was useless material thrown off in the process of growth. Krünitz even thought he had observed that in meadows much visited by bees the plants were more healthy, but the inference he drew was that the honey unless removed was very injurious, that the bees were of use in carrying it off. Sprengel was the first to show that the real office of the honey is to attract insects, but his view was far from meeting with general consent, and even so lately as 1833 were altogether rejected by Kurr who came to the conclusion that the secretion of honey is the result of developmental energy, which afterwards concentrates itself on the ovary.

No doubt, however, seems any longer to exist that Sprengel is right in considering that the object is to attract insects, and thus to secure cross-fertilisation. Thus most of the Rosaceæ are fertilised by insects, and possess nectaries, but, as Delpino has pointed out, the genus Poterium, is anemophilous, or wind-fertilised, and possesses no honey. As also the Maples are almost all fertilised by insects and produce honey, but *Acer negundo* is anemophilous and honeyless. So also among the Polygonaceæ, some species are insect-fertilised and melliferous, while, on the other hand, certain genera, Rumex and Oxyria, have no honey and are fertilised by the wind. At first sight it might appear an objection to this view that some plants secrete honey on other parts than the flowers.

Belt and Delpino have, I think, suggested the true function of these extra floral nectaries. The former of these excellent observers describes a South American species of acacia: this tree, if unprotected, is apt to be stripped of the leaves by a leaf-cutting ant, which uses them, not directly for food, but, according to Mr. Belt, to grow mushrooms on. The acacia, however, bears hollow thorns, while each leaflet produces honey in a crater-formed gland at the base, and a small, sweet, pear-shaped body at the tip. In consequence, it is inhabited by myriads of a small ant, which nests in the hollow thorns, and thus finds meat, drink, and lodging all provided for it. These ants are continually roaming over the plant and constitute a most efficient bodyguard, not only driving off the leaf-cutting ants, but, in Belt's opinion, rendering the leaves less liable to be eaten by herbivorous mammalia. Delpino mentions that on one occasion he was gathering a flower of *Clerodendrons fragrans*, when he was suddenly attacked by a whole army of small ants. M. Boissier also makes the interesting observation that many plants produce honey in some countries and not in others. Thus *Potentilla tormentilla* and *Geum urbanum* give honey in Norway, and none, or hardly any, near Paris. Indeed a careful comparison showed that most plants gave more honey in Norway than at Paris. No doubt, in consequence of this, some plants which are visited by insects in the north are neglected in the south. Thus he observed five species of Hieracum· to be frequented both by bees and humble-bees in Denmark, while near Paris they are never visited by those insects.

M. Boissier found that by watering a plant copiously he could increase the supply of honey ; nay, more, that he could even cause some species to give honey which do not generally do so.

The *pollen* of course, though very useful to insects, is also essential to the flower itself ; but the scent and the honey, at least in their present development, are mainly useful in securing the visits of insects, though the honey is also sometimes of service in causing the pollen to adhere to the proboscis of the insect.

Bees rarely visit flowers unless for some real advantage. The Diptera (flies) appear to be less intelligent, and among fly-flowers we find not only those which attract the insects by honey or pollen, but also trap-flowers, as, for instance, the Arum ; and deceptive flowers, such as *Parnassia*, where five of the stamens terminate in a number of beautiful yellow glands which look like drops of honey, or Stapelia, in which the flowers both in colour and smell resemble decaying meat.

That bees are attracted by, and can distinguish, colours, was no doubt a just inference from the observations on their relation to flowers, but I am not cognisant of any direct evidence on the subject. I thought it therefore worth while to make some experiments ; and a selection from them will be recorded in the forthcoming volume of the Journal of the Linnean Society. I placed slips of glass with honey, on papers of various colours, accustoming different bees to visit special colours, and when they had made a few visits to honey on paper of a particular colour, I found that if the papers were transposed the bees followed the colour.

But if flowers have been modified with reference to the visits of insects, insects also have in some cases been gradually modified, so as to profit by their visits to flowers. This is specially the case with reference to two groups of insects, namely, Bees and Butterflies, which have been specially studied by H. Müller with reference to this point ; and from his works the following facts are mainly taken. Although the whole organisation of the insect might be said to have reference to these relations, still the parts which have been the most profoundly altered are the mouth and the legs. If we are asked why we assume that in these cases the mouth and legs have been modified, the answer is, that they depart greatly from the type found in allied insects, and that between this original type and the most modified examples, various gradations are to be found.

The mouth of an insect is composed of (1) an upper lip (Fig. 11 *a*), (2) an under lip (Fig. 11 *d*) (3) a pair of anterior jaws or mandibles (Fig. 11 *b*), and (4) a pair of posterior jaws or maxillæ (Fig. 11 *c*). These two pairs of jaws work laterally,

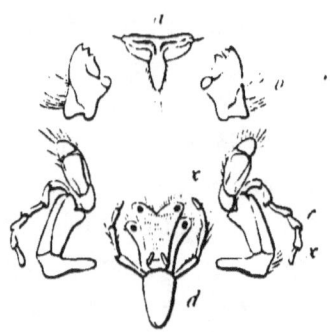

FIG. 11.—Mouth-parts of a Wasp *a*, labrum or upper lip ; *b*, mandibles ; *c*, maxillæ : *d*, labium or lower lip ; *xx*, palpi.

that is to say, from side to side, and not, as in man and other mammalia, upwards and downwards. The lower lip and maxillæ are each provided with a pair of feelers or palpi (Fig. 11, *c* and *d, x*). The above figures represent the mouth parts of a wasp, in which, as is very usually the case, the mandibles are hard and horny, while the maxillæ are delicate and membranous. In the different groups of insects, these organs present almost infinite variations.

Fig. 12 represents the mouth parts of a bee, Prosopis (Fig. 13), seen from below ; *md* being the mandibles ; *pm*, the palpi of the maxillæ *la*, *pl*, those of the lower lip.

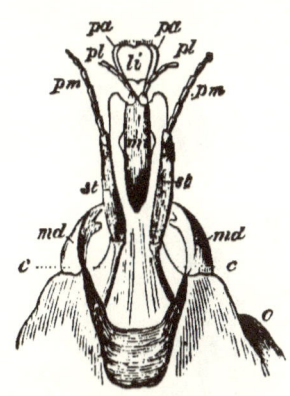

The bees belonging to this genus construct their cells in sand, or in the centre of dry bramble sticks, lining them with a transparent mucus, which they smooth down with their trowel-like lower lip (Fig. 12 *li*), and which hardens into a thin membrane (Smith "Catalogue of Brit. Hymenoptera," p. 7).

FIG. 12.—Front part of head of Prosopis, seen from below, with the mouth-parts extended. *pa*, paraglossæ; *li*. ligula ; *pl*, labial palpi; *pm*. maxillary palpi ; *mt*, mentum; *st*, stipes ; *md*, mandibles ; *c*, cardo ; *o*, eye.

That the mouth of Prosopis probably represents the condition of that of the ancestors of the Hive-bees, before the mouth-parts underwent special modifications, may be inferred from the fact that the same type occurs in allied groups, as is shown in Fig. 14, which represents the mouth of a wasp (Polistes)

FIG. 13.—Prosopis.

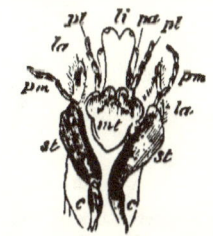

FIG. 14.—Mouth-parts of Polistes.

also seen from below. We may therefore consider that Prosopis shows in this respect no special adaptation for the acquirement of honey, and in fact,

C

though the bees belonging to this genus feed their young on honey and pollen, they can only get the

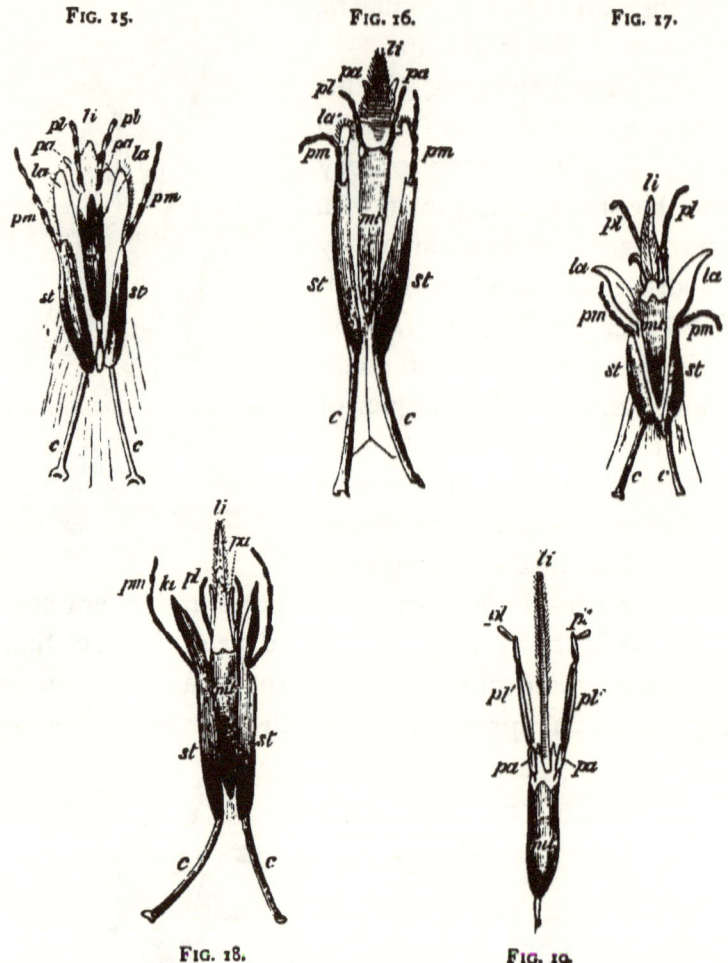

FIG. 15.   FIG. 16.   FIG. 17.

FIG. 18.   FIG. 19.

FIG. 15.—Mouth-parts of Andrena, seen from below—*pa*, paraglossæ; *li*, ligula; *pl*, labial palpi; *pm*, maxillary palpi; *mt*, mentum; *st*, stipes; *c*, cardo; *o*, eye. FIG. 16.—Of Halictus. FIG. 17.—Of Panurgus. FIG. 18.—Of Halictoides. FIG. 19.—Of Chelostoma.

former from those flowers in which it is on the surface. In Andrena (Fig 15), Halictus (Fig. 16),

Panurgus (Fig. 17), Halictoides (Fig. 18), and Chelostoma (Fig. 19), we see various stages in the elongation of the lower lip, until at length it reaches the remarkable and extreme form which it now presents in the hive and humble bees (Fig. 20), and

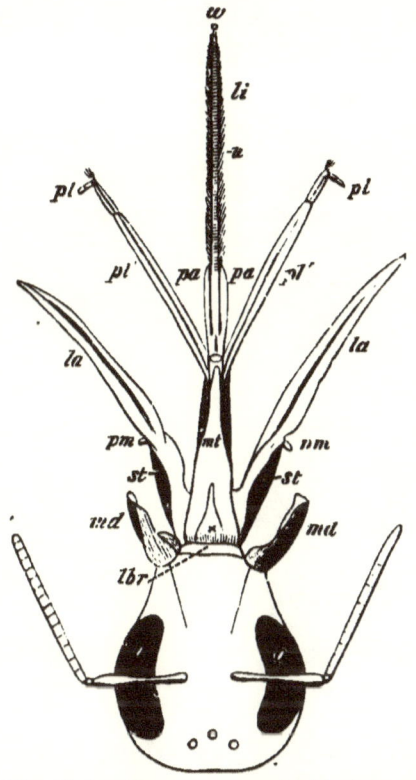

Fig. 20.—Head of Humble-bee (*Bombus agrorum*), with the mouth-parts extended. *pa*, paraglossæ ; *li*, ligula ; *pl*, labial palpi ; *pm*, maxillary palpi ; *la*, lamina of ditto ; *mt*, mentum ; *st*, stipes ; *md*, mandibles ; *c*, cardo ; *o*, eye.

which enables them to extract the honey from almost all our wild flowers. No bees, however, have the proboscis so much elongated as is the case with some butterflies and moths; perhaps, as Hermann Müller has suggested, because the

C 2

necessity of using their mouth for certain domestic purposes has limited its specialisation in this particular direction.

If, again, we examine the hind-legs of bees, we shall find gradations similar to those already mentioned in the lower lip. In Prosopis (Fig. 21) they do not differ materially from those of genera which supply their young with animal food. Portions of the leg, indeed, bear stiff hairs, the original use of

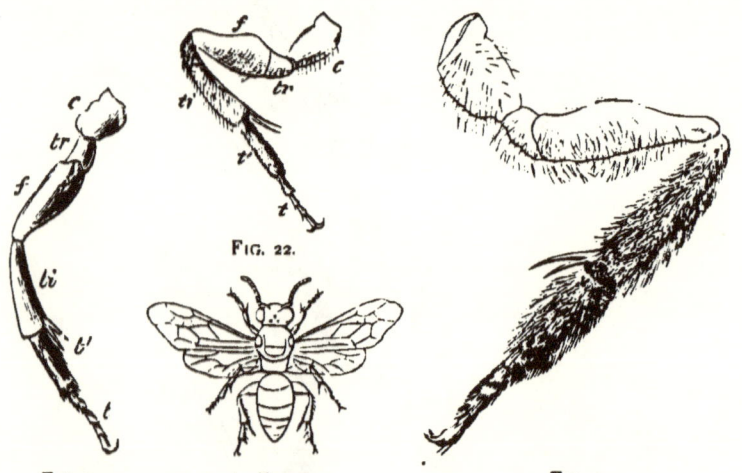

FIG. 22.

FIG. 21.   FIG. 23.   FIG. 24.

FIG. 21.—Left hind-leg of Prosopis.   FIG. 22.—Left hind-leg of Sphecodes.
        FIG. 23.—Sphecodes.   FIG. 24.—Right hind-leg of Halictus.

which, probably, was to clean these burrowing insects from particles of sand and earth, but which in Prosopis assist also in the collection of pollen.

Fig. 22 represents the hind leg of Sphecodes (Fig. 23), a genus in which the tongue resembles in form that of Halictus. Here we see the hairs decidedly more developed, a modification which has advanced still further in Halictus (Fig. 24), in which the de-

velopment of the hairs is most marked on those seg-
ments of the hind legs which are most conveniently
situated for the collection and transport of pollen. In
Panurgus, the same change is still more marked (Fig.
25) and the pollen-bearing apparatus is confined to
the tibia, and first segment of the tarsus, a differentiation which is even more apparent in Anthophora
(Fig. 26). In all these bees the pollen is simply en-
tangled in the hairs of the leg, as in a brush, but there

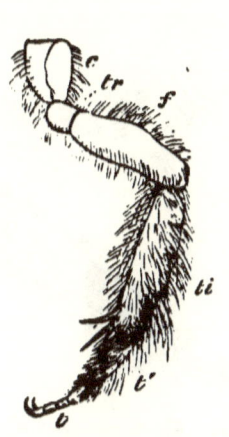

FIG. 25.—Left hind-leg of *Panurgus banksianus.*   FIG. 26.—Right hind-leg of *Anthophora bimaculata.*

are other genera, of which the humble bees and the hive
bees are the only British representatives, which moisten
the pollen with honey, and thus form it into a sticky
mass, which is much more easy to carry, and is borne
not round the leg, but on one side of it. In the
Humble-bee (Bombus, Fig. 27) for instance, the
honey is borne on the outer side of the hinder tibiæ,
which are flattened, smoothed, and bordered by a
row of stiff curved hairs, thus forming a sort of

little basket. Lastly in the Hive-bee (Fig. 28) the
adaptation is still more complete, the hairs on the
first tarsal segment are no longer scattered, but are
arranged in regular rows, and the tibial spurs, inherited
by Bombus from far distant ancestors, have entirely
disappeared.

In some bees the pollen is collected on the body,
and here also we find a remarkable gradation from
Prosopis (Fig. 13) which has only simple hairs like a

F IG. 27.—Right hind-leg of *Bombus*   F IG. 28.—Right hind-leg of Hive-bee.
             Scrimshiranus.

wasp ; through Sphecodes and Nomada, in which the
longer hairs are still few and generally simple (though
some few are feathered) ; to Andrena and Halictus
where the hairs are much more developed ; a change
which is still more marked in Saropoda, Colletes,
and Megachile ; still more so in Osmia and Anthophora ; until we come to the Humble-bees, in
which the whole body is covered with long feathered
hairs.

It is difficult to account for the relations which

exist between flowers and insects, by the hypothesis of a mere blind instinct on the part of the latter.

Thus *Sarcophaga carnaria* visits *Polygonum Bistorta* in search of honey, although that flower does not contain any. *Genista tinctoria* again, though not melliferous, is visited by the males of several species of bees in search of honey. The same is the case with Ononis. H. Müller records a case in which he watched a female Humble-bee (*B. terrestris*) examining an Aquilegia; she made several vain attempts to suck the honey, but after awhile, having apparently satisfied herself that she was unable to do so, bit a hole through the corolla. Having thus secured the honey, she visited several other flowers, biting holes through them, without making any attempt to suck them first; conscious apparently that she was unable to do so He also observed a similar instance in relation to *Primula elatior*. In *Vicia cracca* and some other species, *Bombus terrestris* habitually obtains access to the honey by biting a hole at the base of the flower; and these holes are then subsequently used by other bees. Indeed anyone who has watched bees in greenhouses will see that they are neither confined by original instinct to special flowers, nor do they visit all flowers indifferently.

It would also appear that individual bees differ somewhat in their mode of treating flowers. Some Humble-bees suck the honey of the French bean and Scarlet runner in the legitimate manner, while others cut a hole in the tube and thus reach it, so to say, surreptitiously; and Dr. Ogle has observed that the same bee always proceeded in the same manner some always by the mouth of the flower, others always cutting a hole. He particularly mentions that this was the case with bees of one and the same species, and infers, therefore, that the different individuals differ from one another in their degrees of intelligence ; these observations, though of course not conclusive, are interesting and suggestive.

Lastly, some insects confine themselves to particular flowers. Thus, according to H. Müller—

Andrena florea		Bryonia dioica
Halictoides		Species of Campanula.
Andrena hattorfiana	Visits exclusively.	Scabiosa arvensis.
Cilissa melanura		Lythrum Salicaria.
Macropis labiata		Lysimachia vulgaris.
Osmia adunca		Echium.

The arrangements to which I have hitherto called attention are for the most part of such a nature as to adapt the flowers to the visits of insects. There are others, however, of much interest which serve to protect them from unwelcome visitors, such as ants, who would rob them of their honey without fulfilling any useful purpose in return. Some plants are protected by downward pointing hairs, others by viscid glandular hairs, others by the extreme smoothness of their surface. In other cases the flower is closed by barriers, which only leave sufficient space for the slender proboscis of the bees, while others again, such as the Foxglove, are closed boxes which bees only are able to enter.

Another remarkable peculiarity of plants, which may I think possibly have reference to their relations with insects, is the habit of "sleeping," which characterises certain species.

Many flowers close their petals during rain, which is obviously an advantage, since it prevents the honey and pollen from being spoilt or washed away. Everybody, however, has observed that even in fine weather certain flowers close at particular hours. This habit of going to sleep is surely very curious. Why should flowers do so?

In animals we can understand it; they are tired and require rest. But why should flowers sleep? Why should some flowers do so, and not others? Moreover, different flowers keep different hours. The Daisy opens at sunrise and closes at sunset, whence its name "day's-eye." The Dandelion (*Leontodon taraxacum*) is said to open about seven and close about five; *Arenaria rubra* to be open from nine to three;[1] *Nymphæa alba* from about seven to four; the common Mouse-ear Hawkweed (*Hieracium Pilosella*) from eight to three; the Scarlet Pimpernel (*Anagallis arvensis*) to waken at seven and close soon after two; *Tragopogon pratensis* to open at four in the morning, and close just before twelve, whence its English name, "John go to bed at noon." Farmers' boys in some parts are said to regulate their dinner-time by it. Other flowers, on the contrary, open in the evening.[1]

Now, it is obvious that flowers which are fertilised by night-flying insects would derive no advantage from

---

[1] In my own observations the opening and closing was more gradual and more dependent on the weather than I should have expected from the statements quoted above.

being open by day ; and on the other hand, that those which are fertilised by bees would gain nothing by being open at night. Nay, it would be a distinct disadvantage, because it would render them liable to be robbed of their honey and pollen, by insects which are not capable of fertilising them. I would venture to suggest, then, that the closing of flowers may have reference to the habits of insects, and it may be observed also in support of this that wind-fertilised flowers do not sleep ; and that some of those flowers which attract insects by smell, emit their scent at particular hours ; thus, *Hesperis matronalis* and *Lychnis vespertina* smell in the evening, and *Orchis bifolia* is particularly sweet at night.

Bees appear, moreover, to be skilful in adapting the hour of their visits to the habits of the plants. Thus M. Boissier tells us ("Les Nectaires," p. 166), that he observed some species of Sempervivum (*S. tectorum, S. arachnoideum, S. montanum, S. reflexum,* and *S. maximum*) growing abundantly on rocks, which secreted honey in the morning only. These plants were much frequented by bees up to midday, but quite deserted in the afternoon. He has also observed that some bees which specially frequented Limes and a field of Clover (*Trifolium repens*), went to the former in the early morning, and did not commence visiting the clover until the dew was off. During the height of summer in Provence, the flowers, he tells us, secrete no honey in the heat of the day ; and the bees also remain at home. Mr. Todd even assures us that at Blidah in Algeria the bees during summer do not work after eight in the morning.

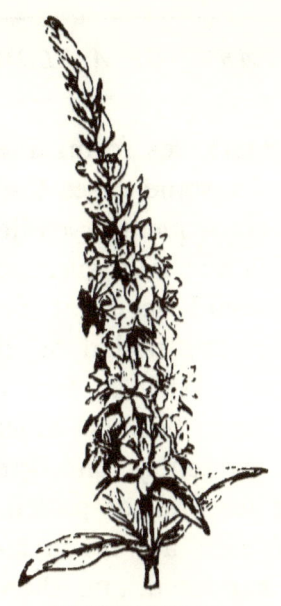

LYTHRUM SALICARIA.

CHAPTER II

I NOW pass to the structure and modifications of flowers. A complete flower consists of (1) an outer envelope or calyx, sometimes tubular, sometimes consisting of separate leaves called *sepals;* (2) an inner envelope or corolla, which is generally more or less coloured, and which, like the calyx, is sometimes tubular, sometimes composed of separate leaves, called *petals;* (3) of one or more stamens, consisting of a stalk or *filament,* and a head or *anther,* in which the pollen is produced; and (4) a pistil or an ovary, which is situated in the centre of the flower, and contains one or more seeds or ovules. The pistil consists of a stalk or *style;* and a *stigma,* to which the pollen must find its way in order to fertilize the flower, and which in many familiar instances forms a small head at the top of the style. In some cases the style is absent, and the stigma is consequently sessile.

Thus, the pistil is normally surrounded by a row of stamens, and it would seem at first sight a very simple matter that the pollen of the latter should fall on the former. This in fact does happen in many instances, and flowers which thus fertilize themselves have evidently one great advantage—few remain sterile for want of pollen.

Such cases, however, are much less frequent than might at first be supposed, and there are three principal modes by which self-fertilization is prevented. Firstly, in many species, the stamens and pistil are situated in different flowers ; such species are called *diclinous ;* when the male and female flowers are on the same plant, they are termed *monœcious ;* when on different plants, *diœcious.*

Secondly, in other instances, as was first discovered by Sprengel, though the stamens and pistil are both situated in one flower, they are not mature at the same time, and the pollen, therefore, cannot fertilize the stigma. These plants are called *dichogamous.* Sometimes, as in the Arum, the pistil matures before the anther, and these plants are called *proterogynous ;* but much more frequently the anther matures before the pistil ; and such plants are called *proterandrous.*

Thirdly, there are some plants in which, as was first discovered by Mr. Darwin, the same object is secured by the existence, within the limits of the same species, of two or more kinds of flowers, differing in the relative position of the stamens and pistil, which are so placed as to favour the transference by insects of the

pollen from the anther of the one form to the pistil of
the other.   These plants are termed *heteromorphous ;*
some of them have two kinds of flowers, and are
called *dimorphous ;* while others have three forms, and
are called *trimorphous.*

But even in plants which belong to none of these
categories we find minor modifications which tend
to prevent self-fertilization ; and Mr. Darwin is probably right in his opinion that no plant invariably
fertilizes itself.   Thus in some species where the
stamens surround the pistil, and which might, therefore, be supposed to be arranged in such a manner as to ensure self-fertilization, the anthers do not
open towards the pistil, but on the sides, and by no
means therefore in a favourable position with reference
to the transference of the pollen.   In most, if not all
the Cruciferæ, the anthers in young flowers have the
side which opens turned towards the pistil ; but before the anthers come to maturity they twist themselves round, so as to turn their backs to the stigma.
Again, in pendent flowers, where the pistil hangs
below the anthers, the stigmatic surface is never the
upper one, which would catch any falling pollen ; but
on the contrary, the lower one, which could hardly be
touched by the pollen of the same flower, but which
is so placed as to come in contact with any insect
or other body approaching the flower from below.

It is also probable that many cases will be found to
exist, in which, though the pollen necessarily comes in
contact with the pistil of the same plant, fertilization
does not take place.   However improbable this might
à priori appear, it is nevertheless said by Hildebrand

to be the case in *Corydalis cava* and *Pulmonaria*
(Fig. 96), by Gärtner in *Verbascum nigrum* (Fig. 98),
and *Lobelia fulgens ;* by Scott in *Primula verticillata,
Oncidium,* &c.

Other cases are recorded in which plants are more
or less insusceptible of fertilization by their own
pollen. Moreover, even where plants are capable of
self-fertilization, the pollen from another flower is
often more effective than their own, whence it follows that if a supply of pollen from another plant
be secured, it is comparatively unimportant to ex-
clude the pollen of the plant itself; for in such cases
the latter is neutralized by the more powerful effect of
the former.

Everyone who has watched flowers, and has observed how assiduously they are visited by insects,
will admit that these insects must often deposit
on the stigma pollen brought from other plants,
generally those of the same species ; for it is a remarkable fact, as Aristotle long ago mentioned, that in
most cases bees confine themselves in each journey to
a single species of plant ; though in the case of some
very nearly allied forms this is not so ; for instance,
it is stated, on good authority, that *Ranunculus acris,
R. repens,* and *R. bulbosus,* are not distinguished by
the bees, or at least are visited indifferently by them,
as is also the case with two of the species of clover,
Trifolium fragiferum and *T. repens.*

Even in the simplest and most regular flowers,
where the stamens surround the pistil, and both are
mature at the same time, insects may visit the flower,
and yet not fertilise it with its own pollen, because

they touch the anther with one side of the proboscis and the stigma with the other. There are, however, in flowers a great many admirable and beautiful contrivances, tending to prevent the fertilization of a flower by its own pollen; in consequence of which insects habitually carry the pollen from the anthers of one flower to the stigma of another.

As already mentioned, there are three principal modes in which self-fertilisation is prevented. Firstly, by the stamens and pistil being situated in different flowers, either on the same plant, or, more commonly, in different plants. These differences form the characteristics of the classes, Monœcia, Diœcia, and Polygamia, of Linnæus ; but it is obvious that such classes are not natural, since we have in very nearly allied species, even within the limits of what is generally considered a single genus, cases in which the one is *diclinous*, that is to say, has the stamens and pistil in separate flowers, while in the other, the flowers contain both.

Secondly, in other cases, the self-fertilization of plants, as was first observed by Sprengel in *Epilobium angustifolium* in the year 1790, is guarded against by the fact that the stamens and pistils do not ripen at the same time.

In some few cases the pistil ripens before the stamens ; these species are called "*proterogynous.*" Thus the Aristolochia has a flower which consists of a long tube with a narrow opening closed by stiff hairs which point backwards, so that it much resembles an ordinary eel-trap. Small flies enter the tube in search of honey, which from the direction of the hairs they can do easily, though on the other hand, from the same cause, it is impossible for them to return. Thus they are imprisoned in the flower ; gradually, however, the pistil passes maturity, and the stigma ceases to be capable of fertilisation, while the stamens ripen and shed their pollen, by which the flies get thoroughly dusted. Then the hairs of the tube shrivel up and release the prisoners, which carry the pollen to another flower.

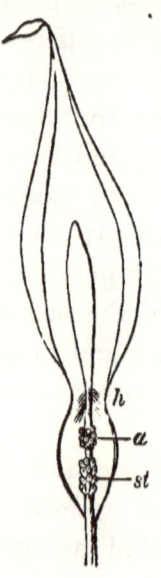

FIG. 29.—Diagrammatic section of Arum. *h*, hairs; *a*, anthers; *st*, stigmas.

Again, in the common Arum, we find a somewhat similar mode of fertilisation. The well-known green leaf, as shown in the annexed diagrammatic figure (Fig. 29), encloses a central pillar which supports a number of stigmas (Fig. 29, *st.*) near the base, and of anthers (*a*) somewhat higher. Now in this case nothing would at first sight seem easier or more natural than that the pollen from the anthers should fall on, and fertilise, the pistils. This, however, is not what occurs. The stigmas mature before the anthers, and by the time the pollen is shed, have become incapable of fertilisation. It is impossible, therefore, that the plant should fertilise itself. Nor can the pollen be carried by wind. When it is shed it drops to the bottom of the tube, where it is so effectually sheltered that nothing short of a hurricane could dislodge it ; and although Arum is common enough, still the chances against any of the pollen so dislodged being blown into the tube of another plant would be immense.

As, however, in Aristolochia, so also in Arum, small insects which, attracted by the showy central spadix, the prospect of shelter or of honey, enter the tube while the stigmas are mature, find themselves imprisoned, by the fringe of hairs (Fig. 29, *h*), which, while permitting their entrance, prevent them from returning. After a while, however, the period of maturity of the stigmas is over, and each secretes a drop of honey, thus repaying the insects for their captivity. The anthers then ripen and shed their pollen, which falls on and adheres to the insects. Then the hairs gradually shrivel up and set the insects free, which carry the pollen with them, so that those which then visit another plant can hardly fail to deposit some of it on the stigmas. Sometimes more than a hundred small flies will be found in a single Arum. In these two cases there is obviously a great advantage in the fact that the stigmas arrive at maturity before the anthers.

Our common *Scrophularia nodosa*, some species of Plantago, &c., are also proterogynous, but such cases are comparatively rare.

The advantage to Scrophularia of being proterogynous, as Mr. Wilson (" Nature," September 5, 1878) has ingeniously pointed out, arises from the fact of its being fertilised by wasps, which generally begin with the upper flower and work downwards, while bees begin below and work upwards. The lower flowers are the older. Hence a bee coming from another plant of the same species fertilises the lower flowers, and then carries off a fresh supply of pollen from the upper and younger ones. On the other hand, as wasps commence from above

D

it is an advantage that the flowers should be proterogynous, because the consequence is that the wasp
fertilises the upper flowers, and then carries off a fresh
supply of pollen from the lower and older ones.

 ˙ On the other hand those in which the anthers come
to maturity before the pistil are much more numerous.
To the category of these plants, which are called
proterandrous, belong some species of Thyme, Pinks,
Epilobium (Figs. 47, 48), Geranium (Fig. 40), Malva
(Figs. 43, 44), (Mallow), Impatiens, Gentians, many
of the Labiatæ, the Umbellifers, most of the
Composites, of the Lobeliaceæ, and Campanulaceæ.
In fact, the greater number of flowers which contain
both stamens and pistil, are more or less proterandrous.

Fig. 30 represents a flower of the Pink in the first,
or male condition. The stamens are mature, and project above the disk of the flower, while the pistil is
still concealed within the tube. On the other hand
Fig. 31 represents the same flower in a more advanced
condition ; the stamens have shrivelled up, while the
pistil now occupies their place.

Again, Fig. 32 represents a flower of the Thyme
(*Thymus serpyllum*) and shows the four mature stamens, *aa*, and the short, as yet undeveloped pistil,
p. Fig. 33, on the contrary, represents a somewhat
older flower, in which the stamens are past maturity,
while the pistil, *p*, on the other hand, is considerably
elongated, and is ready for the reception of the
pollen.

Here it is at once obvious that insects alighting on
the younger (male) flowers would dust themselves with

pollen, some of which, if they subsequently alighted on an older flower, they could not fail to deposit on the stigma.[1] In some cases flowers which are first male and then female, are male on the first day of opening, female on the second. In others the period is longer. Thus Nigella, according to Sprengel, is male for six days, after which the stigma comes to

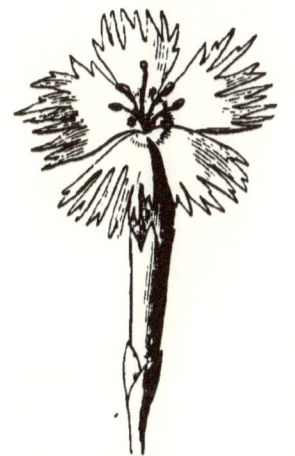

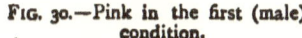

Fig. 30.—Pink in the first (male) condition.

Fig. 31.—Pink in the second condition, with mature stigmas.

maturity and lasts for three or four. (" Das endeckte Geheimniss der Natur," p. 287.)

Fig. 34 represents a flower of *Myosotis versicolor* (a species often known as the Forget-me-not), when just opened. It will be observed that the pistil projects above the corolla and stamens, so that it must be first touched by any insect alighting on the flower. Gradually, however, the corolla elongates, carrying

[1] In the Thymes there are likewise some small flowers which contain no stamens.

up the stamens with it, until at length they come opposite the stigma, as shown in Fig. 35. Thus, if

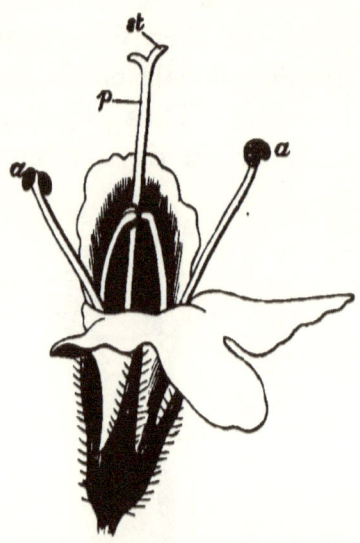

FIG. 32.—*Thymus serpyllum,* in the first condition, with ripe stamens.

FIG. 33.—*Thymus serpyllum,* in the second condition, with mature stigma, *st.*

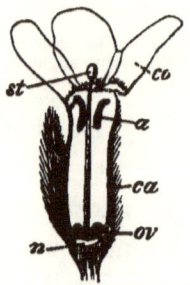

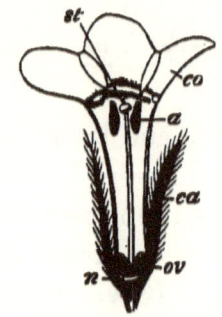

FIG. 34.—*Myosotis versicolor* (young flower).

FIG. 35.—*Myosotis versicolor* (older flower).

the flower has not already been fertilised by insects, it is almost sure to fertilise itself.

I now pass to the third of the principal modes by

which self-fertilisation is prevented. In the flowers
hitherto described, while the several species offer the
most diverse arrangements, we have met with no
differences within the limits of the same species, excepting those dependent upon sex. But there are
other species which possess flowers of two or more
kinds, sometimes, as in the violet, adapted to different conditions, but more frequently so constituted
as to ensure cross-fertilisation. In some of the violets
(*V. odorata, canina*, &c.), besides the blue flowers with

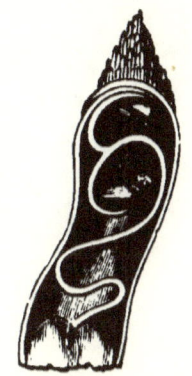

FIG. 36.—Cleistogamous flower of
Lamium amplexicaule.

FIG. 37.—Section of ditto.

which we are all so familiar, there are other, autumnal,
flowers almost without petals and stamens; which
indeed have scarcely the appearance of true flowers,
but in which numerous seeds are produced. "*Cleistogamous*" flowers, as these have been called, occur
also in *Lamium amplexicaule* (Figs. 36 and 37), *Oxalis
acetosella*, *Trifolium subterraneum*, and other plants
belonging to very different groups. They were, I
believe, first observed by Dillenius in Ruellia ("Hortus

Elthamensis," vol. ii. p. 239). As, however, they have no
relation to our present subject, I shall not now dwell
upon them.

I pass on, therefore, to the genus Primula. If a
number of specimens of primroses or of cowslips be
examined, we shall find that about half of them
have the pistil at the top of the tube, and the
stamens half-way down (as is shown in Fig. 38),
while the other half have, on the contrary, the
stamens at the top of the tube, and the pistil half-
way down (as shown in Fig. 39). Corresponding

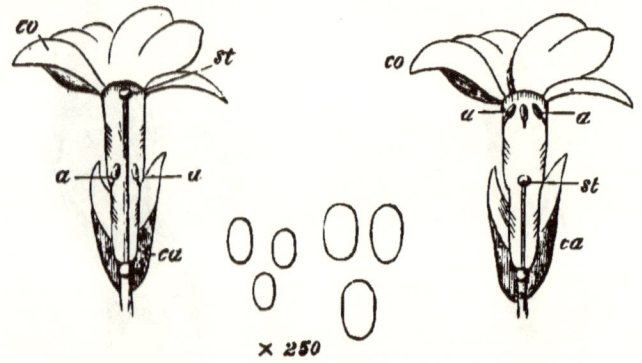

FIG. 38.—Primula (long-styled form).   FIG. 39.—Primula (short-styled form).

differences occur in Polyanthus and Auricula, and
have long been known to gardeners, and even to
schoolchildren, by whom the two forms are distinguished as " thrum-eyed " and " pin-eyed." As
already mentioned, plants which present these differences are known as *Heteromorphous* (in opposition to
those which are *Homomorphous*, or have only one
kind of flower), while those with two forms are called
Dimorphous, those with three, *Trimorphous*.

Sprengel himself had noticed a case of Dimorphism

in Hottonia, and shrewdly observed that there was probably some reason for it, but was unable to suggest any explanation.

In Lythrum the existence of different forms had been observed by Vaucher in 1841, and in the genus Oxalis by Jacquin, who regarded them as indicative of different species; but it was reserved for the genius and perseverance of Mr. Darwin to explain (" Jour. Linn. Soc." 1862, p. 77) the significance of this curious phenomenon, and the important part it plays in the economy of the flower. Now that Mr. Darwin has pointed this out, it is sufficiently obvious : An insect thrusting its proboscis down a primrose of the long-styled form (Fig. 38) would dust its proboscis at a part which, when it visited a short-styled flower (Fig. 39), would come just opposite the head of the pistil, and could not fail to deposit some of the pollen on the stigma. Conversely, an insect visiting a short-styled plant, would dust its proboscis at a part further from the tip ; which, when the insect subsequently visited a long-styled flower, would again come just opposite to the head of the pistil. Hence we see that by this beautiful arrangement, insects must carry the pollen of the long-styled form to the short-styled, and *vice versâ.*

There are other points in which the two forms differ from one another; for instance, the stigma of the long-styled form is globular and rough, while that of the short-styled is smoother, and somewhat depressed. The pollen of the two forms (Figs. 38 and 39) is also dissimilar; that of the long-styled being considerably smaller than the other—$\frac{7}{1000}$ of an inch in diameter against $\frac{10.11}{1000}$, or nearly in the proportion of three to two ; a difference, the importance of which is probably due to the fact that each grain has to give rise to a tube which penetrates the whole length of the style, from the stigma to the base of the flower; and the tube which penetrates the long-styled pistil must therefore be nearly twice as long as in the other. Mr. Darwin has shown that much more seed is set, if pollen from the one form be placed on the pistil of the other, than if the flower be fertilised by pollen of the same form, even taken from a different plant. Nay, what is most remarkable, such unions in Primula are more sterile than crosses between some nearly allied, though distinct species of plants.

The majority of species of the genus Primula appear to be dimorphous, but this is not the case. (Scott, " Proc. Linn. Soc." vol. viii. 1864, p. 80.)

Mr. Darwin has since pointed out (" Jour. Linn. Soc." 1863) that several species of Linum are dimorphous, in the same manner as those of Primula; and has shown that the existence of three forms in Lythrum (Figs. 77—80) already observed by Vaucher, is to be explained in the same manner. I shall refer to this case more in detail when we come to that family. Nor are these by any means the only cases of Heteromorphism now known. I have already mentioned that of Oxalis, and Hildebrand gives the following list of genera as containing Heteromorphous species, viz., Hottonia, Primula, Linum, Lythrum, Pulmonaria, Cinchona, Mitchella, Plantago, Rhamnus, Amsinckia, Mertensia, Leucosmia, Drymospermum,

Menyanthes, and Polygonum. It will be observed that these genera belong to very different groups, while on the other hand, in several cases, as in Primula itself (Scott, "Proc. Linn. Soc." vol. viii.), we find monomorphous and heteromorphous species in the same genus.

Another point of great interest is the spontaneous

FIG. 40.—*Geranium pratense.*

movement of the stamens and pistil in dichogamous plants, first observed by Kolreuter in *Ruta graveolens;* he, however, supposed that the object was to bring the stamens in contact with the pistil ; whereas the real advantage, as Sprengel pointed out, is that in consequence the stamens and pistil successively occupy the same spot in the flower, and thus come in

contact with the same part of the insect. For instance,
in *Geranium pratense* (Fig. 40), when the flower first
opens, the stamens lie on the petals, at right angles
with the upright pistils. As, however, they come to
maturity they raise themselves (Fig. 41 *a*), parallel
and close to the pistil (Fig. 41 *b*), which, however, is
not as yet capable of fertilisation. After they have
shed their pollen, they return to their original position
(Fig. 42), and the stigmas unfurl themselves. More

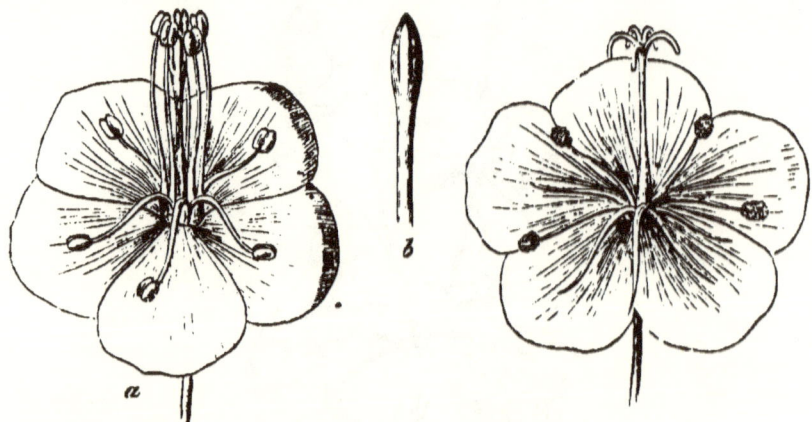

Fig. 41.—*Geranium pratense* (young   Fig. 42.—*Geranium pratense* (older
flower).   Five of the stamens are   flower).   The stamens have retired,
erect.   and the stigmas are expanded.

or less similar movements have been observed in
various other flowers. Thus the cells of the anthers
of the Foxglove (*Digitalis purpurea*) (Figs. 100—102),
which are at first transverse, become longitudinal as
they ripen.

In aquatic plants, the blossoms habitually come to
the surface. In *Valisneria spiralis* the female flower
has a long spiral stalk which enables it to rise to
the top of the water. The male flowers which are

small, very numerous, and attached lower down, separate themselves altogether from the plant, rise to the surface, and fertilise the female flowers, among which they float. When this is effected, the spiral stalk of the female flower again contracts, and draws it down below the surface.

While the *pollen grains* from each flower agree very closely with one another, those of different species differ greatly in form, size, character of surface, &c. Doubtless there are reasons for these differences, but the subject is one with reference to which we have as yet very little information.

According to Sprengel, the pollen of wind-fertilised flowers is drier, and therefore more easily carried by the wind, than that of most insect-fertilised flowers. I say of most, because in some cases, for instance in the violet, as will be shown presently, it is as necessary that the pollen should separate readily from the anthers, as in wind-fertilised flowers.

Mr. Bennett states that the pollen of wind-fertilised flowers is generally spherical ; while that of insect-fertilised flowers is usually furrowed, the furrows running along the longer axis of the grain.

In Dimorphous species the pollen of the short-styled form is generally larger than that of the long-styled form, but in Linum, according to Hildebrand, ("Die Ges. Verth. bei den Pflanzen," p. 37) it is of the same size in both forms.

In Faramea, another Dimorphous group, the surface of the pollen grains is different in the two forms (Thomé "Das Gesetz der vermiedenen Selbstbefruchtung bei den höheren Pflanzen," 1870), the

smaller grains of the long-styled form are studded
with small points; in consequence of which the pollen-grains are less easily detached from the anther;
this difference possibly has reference to the different
position of the two forms; the smooth ones being
sheltered by the flower; while the larger pollen-
grains, which are produced in the anthers of the
long stamens, and are therefore more exposed to the
wind, are, in consequence of their roughness, less
liable to be blown away and wasted.

According to D. Müller ("Bot. Zeit.," 1857) the pollen
of the small flowers of *Viola elatoir* and *V. lancifolia*
is minute and round. Herr von Mohl, however, found
no difference between the pollen of the large and
small flowers in *V. mirabilis* ("Bot. Zeit.," 1863). The
number of grains in these flowers is very small. So
also in the cleistogamous flowers of *Oxalis acetosella*,
there are not above two dozen pollen-grains in each
of the five larger anthers, and one dozen in each of
the five smaller ones. The ovules are about twenty
in number.

It is interesting to notice that the contrivances by
which cross-fertilisation is favoured, or ensured, are
probably of a very different geological antiquity. Thus
as Müller has pointed out, the special peculiarities of
the Umbelliferæ and Compositæ have been inherited
respectively from the ancestral forms of those orders;
those of Delphinium, Aquilegia, Linaria, and Pedicularis, from the ancestral forms of the respective
genera; those of *Polygonum Fagopyrum, P. Bistorta,
Lonicera Caprifolium,* &c., from the ancestors of those
species; while in *Lysimachia vulgaris, Rhinanthus*

Cristagalli, Veronica spicata, Euphrasia Odontites, and *E. officinalis*, we find that differences have arisen even within the limits of one and the same species.

In some species, for instance, we find two varieties, one with larger flowers, which are fertilised by insects, and others with smaller flowers, which are self-fertile.

In other cases the differences between the two kinds of flowers are so marked, and have become so fixed,

FIG. 43.—*Malva sylvestris.*   FIG. 44.—*Malva rotundifolia.*

that the two kinds are usually considered to be distinct species; while in yet other cases the differentiation is still more complete.

Among other obvious evidences that the beauty of flowers is useful to them, in consequence of its attracting insects, we may adduce those cases in which the transference of the pollen is effected in different manners in nearly allied plants, sometimes even in different species belonging to the same genus.

Thus, as Müller has pointed out, *Malva sylvestris* (Fig. 43) and *Malva rotundifolia* (Fig. 44), which grow in the same localities, and therefore must come into competition, are nevertheless nearly equally common. In both species the young flower contains a pyramidal group of stamens which surround the stigma, and produce a large quantity of pollen, which cannot fail to dust any insect visiting the flower for the sake of its honey. In *Malva sylvestris* (Fig. 43), where the branches of the stigma are so arranged (Fig. 45), that the plant cannot fertilise itself, the petals are large and conspicuous, so that the plant is visited by numerous insects ;

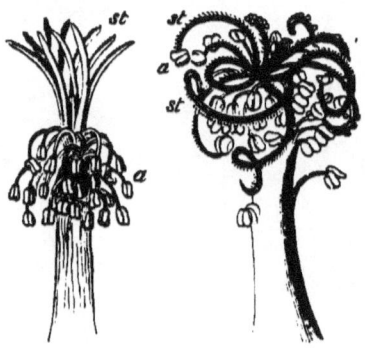

Fiu. 45.—Stamens and stigmas of *Malva sylvestris.*   Fig. 46.—Ditto of *Malva rotundifolia.*

while in *Malva rotundifolia* (Fig. 44), the flowers of which are comparatively small and rarely visited by insects, the branches of the stigma are elongated and twine themselves (Fig. 46) among the stamens, so that the flower can hardly fail to fertilise itself.

Another remarkable instance occurs in the genus Epilobium, which is, moreover, specially interesting, because in *E. angustifolium*, as I have already mentioned, the curious fact was first noticed that the pistil did not mature until the stamens had shed their pollen. *E. angustifolium* (Fig. 47) has conspicuous purplish-red flowers, in long terminal bunches or racemes, and is much frequented by insects ; *E. parviflorum* (Fig. 48), on the contrary, has small solitary flowers, and is seldom visited by insects. Now, to the former species the visits of insects are necessary, since the stamens ripen before the pistil, and the flower has consequently lost the power of self-fertilisation. In the latter, on the contrary, the stamens and pistil come to maturity at the same time, and the flower can therefore

Fig. 47.—*Epilobium angustifolium.*   Fig. 48.—*Epilobium parviflorum.*

fertilise itself. It is, however, no doubt sometimes crossed by the agency of insects ; and indeed I am disposed to believe that this is true of all the flowers which are either coloured or sweet scented.

The genus Geranium also affords us an instructive example. There are a number of species which, as will be seen in Fig. 49, differ much in the size of the flowers. Thus those (Fig. 49 *a*) of *Geranium pratense*

(Fig. 40) are nearly twice as large as those of *G. pyre-naicum* (Fig. 49 *b*), which again are much larger than those of *G. molle* (Fig. 49 *c*), while those of *G. pusillum* (Fig. 49 *d*) are still smaller. These differences of size appear to be connected with other remarkable differences between these species. Fig. 41, as already mentioned, represents a flower of *G. pratense* when first opened. Five of the stamens have raised themselves and stand upright, and surround the still immature pistil. When they have shed their pollen they sink back and shrivel up, when the other five raise themselves. At a later stage these in their turn fall back and shrivel up, but the stigma does not become mature (Fig. 42), until all the stamens have shed their pollen. Under these circumstances *G. pratense* has lost the power of self-fertilisation, and is absolutely dependent on the visits of insects.

FIG. 49.—Corolla of—*a*, *Geranium pratense; b, G. pyrenaicum; c, G. molle; d, G. pusillum.*

G. *pyrenaicum* (Fig. 49 *b*) is also pro-terandrous; but while in *G. pratense* the pistil is not mature until the stamens have shed all their pollen and fallen back, in *G. pyrenaicum* the second series of stamens are still upright when the stigmatic lobes unfurl; the flower is consequently less absolutely dependent on insects, and we see that the corolla is much smaller.

In the third species, *G. molle* (Fig. 49 *c*), the pistil matures before the second series of stamens, and the corolla is still smaller; while in *G. pusillum* (Fig. 49 *d*) the pistil matures before any of the stamens. Thus then these four species may be arranged in a table as below :—

GERANIUM PRATENSE.	GERANIUM PYRENAICUM.	GERANIUM MOLLE.	GERANIUM PUSILLUM.
Flower large.	Flower small.	Flower smaller.	Flower smallest.
First exclusively male, then exclusively female.	First exclusively male, then hermaphrodite.	First exclusively male, then hermaphrodite.	First exclusively female, soon becoming hermaphrodite.
Incapable of self-fertilization.	Generally fertilised by insects.	Often self-fertilised.	Generally self-fertilised.

Indeed, though further observations on the point are no doubt required, it would seem that, as a general rule, where we find within the limits of one genus some species which are much more conspicuous than others, we may suspect that they are also more dependent on the visits of insects.

Sprengel also suggests, and, as it would appear, with reason, that the lines and bands by which so many flowers are ornamented have reference to the position of the honey ;[1] and it may be observed that these honey-guides are absent in night-flowers, where of course they would not be visible, and would therefore be useless, as, for instance, in *Lychnis vespertina* (Fig. 50), or *Silene nutans.* Night-flowers, moreover,

---

[1] I did not realise the importance of these guiding marks until, by experiments on bees, I saw how much time they lose if honey, which is put out for them, is moved even slightly from its usual place.

E

are generally pale ; for instance, *Lychnis vespertina* is white, while *Lychnis diurna*, which flowers by day, is red. Brown flowers, such as Scrophularia, some species of Epipactis, of Lonicera, &c., perhaps owe their hue to the selective influence of wasps. Fly flowers also are often livid or flesh-coloured.

I have been good-humouredly accused of attacking the Bee, because I have ventured to suggest that she

FIG. 50.—*Lychnis vespertina.*

does not possess all the high qualities which have been popularly and poetically ascribed to her. But if scientific observations do not altogether support the moral and intellectual eminence which has been ascribed to Bees, they have made known to us in the economy of the hive many curious peculiarities which no poet had dreamt of, and have shown that bees and other insects have an importance as regards flowers which had been previously unsuspected. To

them we owe the beauty of our gardens, the sweet-
ness of our fields.   To them flowers are indebted for
their scent and colour ; nay, for their very existence
in its present form.   Not only have the present shape
and outlines, the brilliant colours, the sweet scent
and the honey of flowers, been gradually developed
through the unconscious selection exercised by insects ;
but the very arrangement of the colours, the circular
bands and radiating lines, the form, size, and position
of the petals, the relative situations of the stamens
and pistil, are all arranged with reference to the
visits of insects, and in such a manner as to ensure
the grand object which these visits are destined to
effect.

LYCHNIS VESPERTINA.

CHAPTER III.

DICOTYLEDONS.

THALAMIFLORÆ.

IN the preceding chapters I have endeavoured to give
a general sketch of the relations existing between
flowers and insects. I shall now proceed to de-
scribe particular instances more in detail, following in
general the classification adopted in Mr. Bentham's
admirable " Handbook of the British Flora," from
which also many of my facts and illustrations have
been borrowed. I propose to go through the English
Flora, in the order of Mr. Bentham's work, calling attention to those facts, bearing on our present subject,
which strike me as most interesting. The present
chapter is devoted to the thalamifloral division of the
Dicotyledons.

The vegetable kingdom may be divided into flowering and flowerless plants; while flowering plants again fall into two divisions, known as Dicotyledons or Exogens and Monocotyledons or Endogens. Dicotyledonous or exogenous plants are those in which, when the seed germinates, the "*plumule*" or bud arises between two (rarely more) seed-leaves or *cotyledons* of the embryo, or from a terminal notch. In this class the leaves have their nerves branched, forming a sort of network, as in the oak, beech, clover, violet, &c.   In their growth they increase by forming new woody tissue over the old, whence the term "Exogenous."   In a Dicotyledonous or exogenous tree, therefore, we find a number of concentric circles, each representing a period of growth, and indicating, though roughly, its age in years. Monocotyledonous or endogenous plants, on the contrary, are those in which the plumule or bud is developed from a sheath-like cavity on one side of the cotyledon.   The leaves have parallel nerves, as for instance in grasses, orchids, lilies, palms, &c.   In a cross-section the wood shows no concentric circles, but consists of bundles of woody fibre irregularly imbedded in cellular tissue.   Both these classes have flowers.

Cryptogams, on the contrary (ferns, mosses, seaweeds, lichens, fungi, &c.), have no flowers, and multiply by bodies called spores.

That the colour of the corolla has reference to the visits of insects is also well shown by the case of those flowers, which—as, for instance, the ray or outside florets of Centaurea—have neither stamens nor pistils, and merely serve, therefore, to render the flower-head more conspicuous. The calyx, moreover, is usually green ; but when the position of the flower is such that it is much exposed, it becomes brightly coloured, as, for instance, in the Berberry or Larkspur

The above characters, though true in the main, do not hold good in all cases. For instance the genus Arum, though a Monocotyledon, has reticulated nerves, but its stem is endogenous, and its embryo has only one cotyledon.

The class of Dicotyledons is divisible into four sub-classes, which may be thus characterised :—

> *Thalamiflora.* Petals distinct from the calyx and from each other, seldom wanting. Stamens usually hypogynous (*i.e.* attached under the ovary), so that if the calyx be torn away the stamens remain.

> *Calyciflora.* Petals usually distinct. Stamens perigynous (*i.e.* attached round the ovary), or epigynous (*i.e.* placed upon the ovary).

> *Corolliflora* or *Monopetala.* Petals united (at least, at the base) into a single corolla.

> *Incompleta* or *Monochlamydea.* Perianth or floral envelope, really or apparently simple ; or none.

These subclasses may be tabulated as follows :—

Perianth or floral envelope
- single or none . . . . Monochlamydeæ.
- double
  - corolla of united petals . . . . Corollifloræ.
  - corolla of distinct petals
    - stamens hypogynous . . } Thalamifloræ.
    - stamens perigynous or epigynous . . . } Calycifloræ.

RANUNCULACEÆ.

This order contains fourteen British genera, including the Clematis, Ranunculus (Buttercup), Anemone, Columbine, Hellebore, Larkspur, Pæony, &c.

In the Buttercup (*Ranunculus acris*), the anthers commence to discharge their pollen, as soon as the flower opens, beginning from the outside. The stigmas, however, are not as yet mature, nor do the stamens open on the side which is turned towards them, but on the contrary, on their edges; moreover as each stamen ripens, it generally turns outwards. The result of this is that bees and other insects, which visit the flowers in search of honey, are almost sure to dust themselves with pollen; which they carry away with them, and are then very likely to deposit it on another flower. The stigmas are mature before the inner stamens have shed all their pollen, and self-fertilisation must often take place, both by means of the small insects which may almost invariably be found wandering about the flower, and because the inner stamens often touch some of the stigmas. Larger insects, however, which fly from flower to flower, must habitually carry the pollen from the younger flowers, and deposit it on the stigmas of those more advanced.

Clematis recta produces no honey, but is visited for the sake of the pollen. It is proterandrous (see p. 28), but not very decidedly so; for as in other flowers which do not produce honey, if the stamens had shed all their pollen before the pistil came to maturity, insects would cease to visit the flowers before the stigma had attained maturity, and had thus become susceptible of fertilisation.

Like Clematis, *Thalictrum* produces no honey. The petals are absent and the sepals minute, but the stamens are numerous and brightly coloured.

Caltha palustris has large yellow sepals, but no true petals. In the *Hellebore* also the petals are minute, but secrete honey. The species of this genus are said by Hildebrand to be proterogynous. (See p. 28.)

In *Anemone nemorosa* the colouring is given not by the corolla, but by the calyx. The flower does not appear to produce honey, but bees are said to pierce the base of the flower, and lick the sap. Van Tieghem however states that it gives off honey from the whole surface of the receptacle.

Delphinium (the Larkspur) and *D. elatum* (Figs. 51 —54) have been well described by H. Müller. The five sepals are brightly coloured ; the upper one is produced into a long spur (*x x*). The two upper petals are also produced into spurs which lie within the former, and secrete honey. In order to reach this it is necessary for the bee to press its proboscis between the upper and lower petals, through the interval (Figs. 51, 53 *m*). The lower wall of this orifice is in front closed by the lower petals (Figs. 51, 53 *pe pe*), which are turned upwards and sideways, so as to form the lower wall of the orifice leading to the nectary, and to cover the stamens and pistils. Immediately behind the entrance to the tube, however, these petals contract so as to leave a space (*m*). The stamens (*a*) and pistil lie below this space, and as the stamens ripen, they successively raise themselves and their anthers pass through this space, as shown in

Fig. 51 *a'*, so that the proboscis of the bee, in passing
down to the honey can hardly fail to come in contact
with them. After shedding their pollen, they turn
down again, and when each anther has thus raised
itself and again retired, the pistil in its turn takes pos-

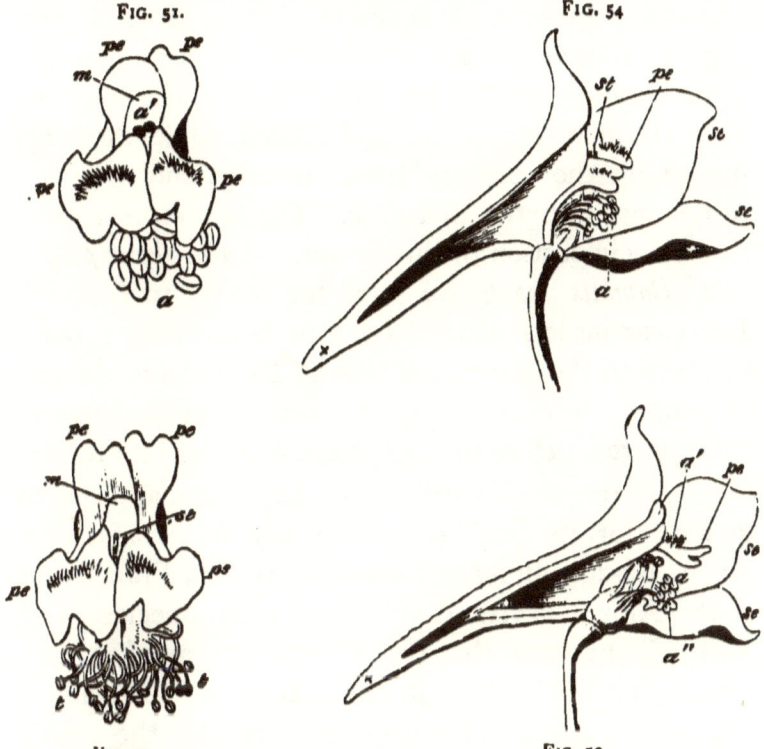

FIG. 51.   FIG. 54.

FIG. 53.   FIG. 52.

FIG. 51.—A young flower of *Delphinium elatum*, seen from the front, and after re-
moval of the calyx.
FIG. 52.—Section of the same flower seen from the side.
FIG. 53.—An older flower, seen from the front, after removing the calyx.
FIG. 54.—Section of the same flower, seen from the side.

session of the place, as shown in Fig. 53, and 54 *st*;
and is thus so placed, that a bee which has visited
a younger flower and there dusted its proboscis, can
hardly fail to deposit some of the pollen on the

stigma. Fig 51 represents a young flower seen
from the front, and after the removal of the calyx ;
it shows the entrance leading to the nectary, in
which are seen the heads of two mature stamens, *a'*,
while the others, *a a*, are situated in a cluster below.
Fig. 52 represents a section of the same flower.
Fig. 53 represents a somewhat older flower, in the
same position as Fig. 51. In this case the stamens
have all shed their pollen and retired, while the stigmas *st*, on the contrary, have risen up, and are seen
projecting into the space *m*. Fig. 54 represents a
side view in section of this flower. *Anthophora pilipes*
and *Bombus hortorum* are the only two North
European insects, which have a proboscis long enough
to reach to the end of the spur of Delphinium elatum.
A. pilipes, however, is a spring insect, and has already
disappeared, before the Delphinium comes into flower,
which, in the neighbourhood of Lippstatt, appears to
depend for its fertilisation entirely on *Bombus hortorum*, though Boissier assures us that in France and
in the Alps it is visited by several other species.

It will be seen that the Ranunculaceæ offer very
remarkable differences in the manner of their adaptation to insects. Honey is secreted by the sepals
in certain Pæonies ; by the petals in Ranunculus,
Delphinium, Helleborus, &c. ; by the stamens in Pulsatilla ; by the ovary in Caltha ; while it is entirely
absent in Clematis, Anemone, and Thalictrum. The
conspicuousness of the flower is due to the corolla in
Ranunculus ; to the calyx in Anemone, Caltha, and
Helleborus ; to both in Aquilegia and Delphinium ;
to the stamens in Thalictrum. The honey is in some
cases easily accessible, in others it is situated at the

end of a long spur. The former species are capable of self-fertilisation, the latter are said by H. Müller to have lost their power.

BERBERIDEÆ.

The common Berberry is the only British representative of this order, though *Epimedium alpinum* has by some been considered to be indigenous ; as Mr. Bentham thinks, on insufficient grounds.

In the common Berberry (*Berberis vulgaris*), the stamens (Figs. 55 *ff*, 56 *a*) lie close to the petals and almost at right angles to the pistil, as shown in

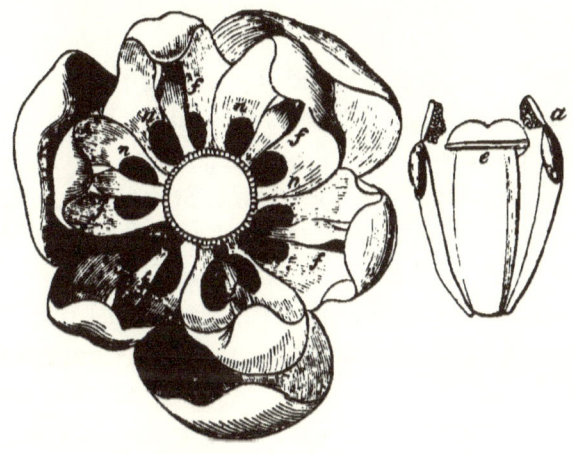

FIG. 55.   FIG. 56.

FIG. 55.—Flower seen from above.
FIG. 56.—Pistil with two stamens, after the visit of an insect.

Fig. 55. The honey-glands (*n n*) are twelve in number, situated in pairs at the base of the petals, so that the honey occupies the angle between the bases of the stamens and of the pistil. The papillary edge of the summit of the pistil (*e*) is the stigma. In open flowers of this kind it is of course obvious that insects will dust themselves with the pollen and then carry it with them to other flowers. In Berberis, however, both advantages, the dusting and the cross-fertilisation, are promoted by a very curious contrivance. The bases of the stamens are highly irritable, and when an insect touches them the stamens spring forward to the position shown in Fig. 56 and strike the insect. The effect of this is not only to shed the pollen over the insect, but also in some cases to startle it and drive it away, so that it carries the pollen, thus acquired, to another flower.

NYMPHÆACEÆ.

This order is represented by two British species. *Nymphæa alba*, the White Waterlily; and *Nuphar lutea*, the Yellow Waterlily. According to Delpino, *N. alba* is fertilised by beetles. Sprengel contrasts the large size of the pistil and the great number of the stamens in *N. lutea*, where the fertilisation is, as it were, a matter of accident, with the small pistil and four stamens of a Labiate ; such, for instance, as the common Dead Nettle, which, as we shall see, are so beautifully arranged with reference to one another, and where consequently so much less pollen is required.

PAPAVERACEÆ.

Of this family the Common Poppy is the best known representative, though the Celandine is also common on roadsides, especially near villages. The Poppy has two sepals, which drop off as the flower expands ; four petals ; numerous stamens, forming a ring round a globular or ovoid pistil, which is crowned by a circular disk, on which the stigmas radiate from the centre. The flowers secrete no honey, but are visited for the sake of the pollen. Owing to the weakness of the petals, insects naturally alight on the stigma, which forms a most convenient stage for them in the centre of the stamens, and they thus naturally carry the pollen from one flower to another.

FUMARIACEÆ.

This natural order contains only two British genera, Fumaria and Corydalis. The flowers of Fumaria have not yet, I think, been satisfactorily explained. Their form and arrangement are very singular, but they are not very conspicuous, and are said to be little visited by insects, being, according to Müller, self-fertile.

In Corydalis, on the contrary, the flowers are much larger, more conspicuous, and, at least in *C. cava*, are said to have lost the power of self-fertilisation. Hildebrand has found (Ueber die Bestaübungs Vorrichtungen bei den Fumariaceen) that they are absolutely sterile with their own pollen, and only imperfectly fertile with that from other flowers of the same plant, so that they can only be completely fertilised by that from a different plant. The tube of the flower is 12 millimetres long, and as the honey only occupies at most 4—5 millimetres, it is inaccessible to the Hive bee, whose proboscis is only 6 mm. long, and almost so to the common humble bee, in which it is 7—9, or at most 10 mm. long. The latter can reach the honey, but not lap it conveniently. She however, is in the habit of biting a hole through the tube, by which means she obtains access to the honey, and in some plants the greater number of flowers will be found to have been treated in this manner. Several other bees, for instance, the hive bee, *Andrena albicans*, K.; *A. nitida*, Fourc.; *Sphecodes gibbus*, L.; and *Nomada fabricana*, L., have been observed by Müller to make use of the entrance thus prepared for them. Moreover, though the hive bees are unable to suck the flowers in their natural condition, the flowers are visited by them for the sake of their pollen.

The upper petal is produced into a long spur. The two middle petals form a sort of sheath, sur-rounding the stamens and pistil; at about a third of their length from the base is a peculiar fold of the edge, which acts as a sort of hinge, so that the terminal part, which forms a sort of sheath or cap to the anthers and stigma, is somewhat moveable. The stamens are united into an upper and lower group. The upper basal edge of the upper group is produced into a long spur, which lies in the spur of the upper petal, and the tip of which secretes honey. When a bee visits the flower, she depresses the anther cap, and the anthers and pistil thus exposed rub against her breast. When the pressure is removed the cap resumes its place and again protects the anthers and pistil. Our common English *Fumaria officinalis* is formed on the same plan as *Corydalis cava*, the spur, however, being quite short. It appears, moreover, to be self-fertile, and in spite of its complex organisation seems to be but rarely visited, at least by day.

Hildebrand never saw an insect on the flowers. H. Müller saw them occasionally visited by the honey bee. In *F. officinalis,* as in *C. cava,* the anther cap is elastic, and on the departure of the insect resumes its original place. It is interesting that in other species of each genus (none of which however are English), as for instance *C. ochroleuca* and *F. spicata,* the pillar formed by the stamens and pistil is in a state of tension, but is retained in its place by the two petals forming the cap. These are as it were locked together, but when once separated by the pressure of the bee, the pillar formed by the stamens and pistil is set free, and springs up, thus dusting the insect. This process only happens once in each flower. Though these species are not British, I mention this here, because we shall find a very similar process in some of the Leguminosæ (p. 86), and it is most interesting to find such a remarkable arrangement thus repeated in very different groups.

CRUCIFERÆ.

The Wallflower, Stock, Cabbage, Shepherd's Purse, Watercress, &c., belong to this group.

The Cruciferæ are easily distinguished from other orders by their four sepals and petals, and six stamens; but the genera into which they are divided are by no means so well marked, and are to a great extent distinguished by differences in the pods and seeds. The general structure of the flower is more or less similar throughout the order, but the number and position of the honey-glands differ in almost every species. *Hesperis matronalis* is one of those plants which are specially odoriferous in the evenings, and is therefore probably in most cases fertilised by moths, though it is also visited by day-insects, as for instance, by the hive bee, the white butterflies (*Pieris brassicæ, P. rapi,* and *P. napi*), *Halictus leucopus, Andrena albicans, Volucella pellucens, Rhingia rostrata,* &c.

But though the colour, honey, and scent of the Cruciferæ have evident reference to the visits of insects, this order does not offer so many special and specific adaptations as we shall meet with in other groups ; and the majority of species, at any rate, appear to have retained the power of self-fertilisation.

RESEDACEÆ.

Flowers bisexual, small, greenish, sometimes scented irregular. Sepals and petals 4-7. Stamens many inserted on a broad disk. Pistil one, with 2-3 stigmas.

This order is represented in Britain by one genus Reseda (the Mignonette), containing three species. In the common garden mignonette the upper half of the base of the flower raises itself between the stamens and the sepals into a quadrangular, perpendicular plate, which is first yellowish, and after the flower has faded, brown. It is enclosed in a sort of box, the three upper petals forming the lid. Its hinder surface secretes honey. The mignonette is said to be specially frequented by bees of the genus *Prosopis.*

CISTINEÆ.

This order contains only a single British genus, Helianthemum, with four species. The flowers do not secrete honey. The stamens are numerous. As the pistil projects above them, insects, in alighting on the flower, generally touch the pistil before the stamens ; and cross-fertilisation must therefore often take place. At the same time, if from any cause insect-visits are deferred, the flower is almost sure to fertilise itself.

VIOLACEÆ.

This order is limited in Europe to the single genus Viola, of which we · have, according to Bentham, five English species. Besides the showy, coloured flowers with which we are all familiar, most of the species possess minute flowers, which, however, produce abundance of seed. These appear later in the year, and are not only much smaller than the others, but almost without petals. In fact, according to Bentham, the Pansy (*V. tricolor*) is the only one of our English species in which the showy flowers generally produce seed. The presence of these two totally different kinds of flowers is a very interesting fact ; and as the smaller, or as they are called, "*cleisto-gamous*" flowers are sufficient to reproduce the species, and of course have the advantage of requiring much less expenditure of material, the persistence of the showy ones can only, I think, be accounted for by the fact that the ordinary flowers are useful in securing an occasional cross, as the cleistogamous flowers habitually fertilise themselves.

Viola canina. The structure of the coloured flowers is very curious, and has been well described by Sprengel. The petals are five in number, and irregular in form ; the median one being produced into a

F

hollow spur (Fig. 57 *f*), the entrance to which is protected partly by the stigma, partly by two tufts of hairs, or rather of delicate lobular processes, situated on the two median petals. The stamens consist of a short filament, to which the anther is attached, and a terminal membranous expansion, while the two

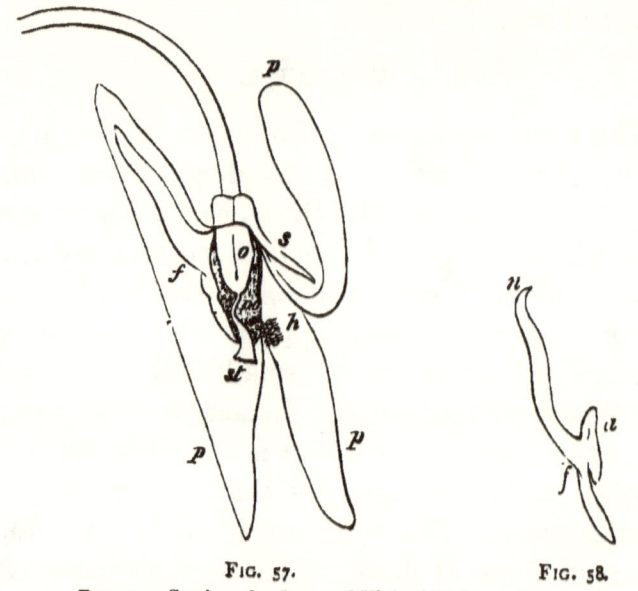

Fig. 57.

Fig. 58.

Fig. 57.—Section of a flower of Violet (*Viola canina*).
Fig. 58.—Stamen of ditto.

lower stamens also send out each a long spur (Fig. 58 *n*), which lies within the spur of the median petal, and secretes honey at its fleshy end. The terminal membranous expansions of the five stamens slightly overlap one another, and their points touch the pistil, so that they enclose a hollow space. The pollen differs from that of most insect-fertilised flowers, in being drier, and more easily detached from the anthers; consequently, when the latter open, the pollen drops out ; and as the flower is reversed and hangs down, the pollen falls into the closed space between the pistil and the membranous terminations of the stamens. The pistil is peculiar, the base of the style not being straight as usual, but thin and bent (Fig. 57). The stigma *st* is the enlarged end of the pistil ; and shows several small fleshy projections. It will be obvious from the above description that when a bee visits the flower, her head will come in contact with and shake the stigma, thus opening, as it were, the box containing the pollen, and allowing it to fall on the head of the bee. It is thus carried away, and some can hardly fail to be deposited on the stigma of the next violet which the bee visits.

Sprengel, in his description of *V. odorata*, gives the following list of questions and answers as regards this species ; passing over, however, the more general points, such as the secretion of honey, the colour of the corolla, the radiating lines on the petals, and the smell.

1. Why is the flower situated on a long stalk, which is upright, but curved downwards at the free end ?—In order that it may hang down ; which, firstly, prevents rain from obtaining access to the honey ; and, secondly, places the stamens in such a position that the pollen falls into the open space between the pistil and the free ends of the stamens. If the flower were upright, the pollen would fall into the space between the base of the stamen and the base of the pistil, and would not come in contact with the bee.

2. Why does the pollen differ from that of most other insect-fertilised flowers ?—In most of such flowers the insects themselves remove the pollen

from the anthers; and it is therefore important that the pollen should not easily be detached and carried away by the wind. In the present case, on the contrary, it is desirable that it should be looser and drier, so that it may easily fall into the space between the stamens and the pistil. If it remained attached to the anther, it would not be touched by the bee, and the flower would remain unfertilised.

3. Why is the base of the style so thin?—In order that the bee may be more easily able to bend the style.

4. Why is the base of the style bent?—For the same reason. The result of the curvature is that the pistil is much more easily bent than would be the case if the style were straight.

5. Finally, why does the membranous termination of the upper filament overlap the corresponding portions of the two middle stamens?—Because this enables the bee to move the pistil, and thereby to set free the pollen more easily than would be the case under the reverse arrangement.

In *Viola tricolor*, the form of the stigma is very different from that of *V. canina*, but the reason of the difference has not been satisfactorily explained. Mr. Bennett considers that this species is fertilised by Thrips. Mr. Darwin, however, has satisfied himself that when bees are excluded, it is comparatively infertile, and he has favoured me with the following memorandum on the subject.

"When," he says, " I formerly covered up a fine, large, cultivated variety, it set only 18 capsules, and most of them contained very few good seeds, several from only 1 to 3 ; whereas an equally fine uncovered plant, growing close by, produced 105 fine capsules.

The few capsules which are produced when insects are excluded are probably due to the curling up of the petals (as Fermond and F. Müller remark) as they wither, by which process pollen-grains adhering to the papillæ may be inserted into the cavity of the stigma. The moth Plusia is said to visit the flowers largely. Humble-bees are common agents in fertilising these flowers; but I have seen more than once a fly (*Rhyngia rostrata*) with the under side of its body, head, and legs dusted with the pollen of this plant, and having marked the flowers which they had visited, found, after a few days, that they had all been fertilised.

" It is curious in this case, as in many others, how long the flowers may be watched without seeing one visited by an insect. During one summer, I repeatedly watched some large clumps of heartsease, many times daily, for a fortnight, before I saw a humble-bee at work. During another summer I did the same, and then one day, as well as on two succeeding days, I saw a dark-coloured humble-bee visiting almost every flower in several clumps; and after a few days almost all the flowers suddenly withered, and produced fine capsules. A certain state of the atmosphere seems to be necessary for the secretion of nectar, and as soon as this occurs, it is perceived by various insects, I presume by the odour emitted by the flowers, and these are immediately visited."

POLYGALACEÆ.

This order contains, according to Bentham, but one British species, which, however, is very common, the

Milkwort (*Polygala vulgaris*), Fig. 59. The structure
of the flower is curious, and was first explained by
Hildebrand, whose account, however, does not seem
to me entirely complete or satisfactory. There are
five sepals (Figs. 60, 61 *s s*), of which three are small,
linear, and greenish; the other two much larger,
coloured like the petals, and obovate or oblong. The

Fig. 59. – *Polygala vulgaris.*

petals form a tube to the inside of which the stamens
are attached in two bundles (Fig. 61 *a*), and which
contains a number of white hairs pointing downwards,
while near the upper end are two groups of finger-
like lobes. The pistil (Fig. 61 *st*) occupies the axis of
the flower, and ends in a spoon-shaped hollow. The
short stamens lie just over this hollow, and shed their
pollen into it, after which they withdraw a little to the
side. Close behind the hollow is a projection which

terminates in a very viscid disk. When the proboscis of an insect is forced down the tube in search of honey, it comes in contact with this viscid disk, and being thus rendered adhesive, when it is withdrawn carries some of the pollen with it, and thus conveys it to the next flower, where it is stripped off the retreating proboscis by the edge of the viscid disk,

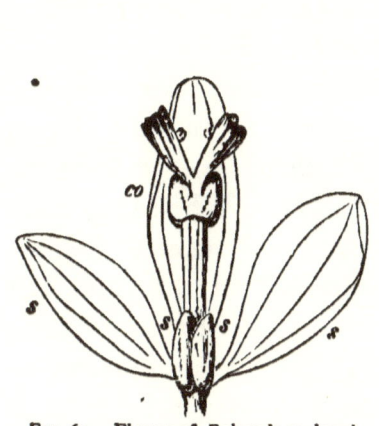
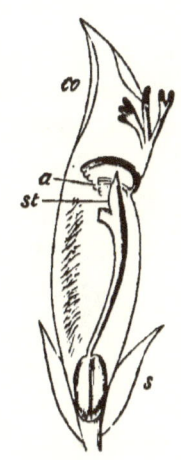

FIG. 60.—Flower of *Polygala vulgaris.*   FIG. 61.—Section of ditto.

and is thus accumulated in the stigmatic hollow. *Polygala vulgaris* is sometimes blue and sometimes pink ; why is this ? It is, moreover, a variable species in other respects, as for instance in the size and proportions of the different leaves. The use of the curious finger-formed processes has not, I think, been satisfactorily explained.

CARYOPHYLLACEÆ.

This is a large family and contains fourteen British genera ; Dianthus (the wild Pink), Saponaria, Silene, Lychnis (Fig. 50), Sagina, Cherleria, Arenaria, Mæn-

chia, Holosteum, Cerastium, Stellaria (Fig. 62), Sper-
gularia, Spergula, and Polycarpon.

In Dianthus, of which we may take *D. deltoides*,
the Maiden Pink, as an illustration, the stamens
are united with the petals at the base, and form a
yellow, fleshy, swelling which secretes honey. The
tube of the flower is so narrow, and so nearly closed
by the stamens and pistil, that the proboscis of
Lepidoptera alone can reach the honey, though flies
and other insects visit it for the pollen. The upper
surface of the flower forms a flat disk, pink or spotted
with white. The stamens are ten in number. Soon
after the flower opens, five of them emerge from the
tube, ripen, and the anthers open. When they have
shed their pollen, the other five do the same. During
this period the pistil is concealed in the tube, but
after the anthers have ripened and shed most of their
pollen, it also emerges and the two long stigmas
expand themselves. These two stages have been
already referred to (see Figs. 30 and 31). Under
these circumstances the butterflies can hardly fail to
carry the pollen from the anthers of young flowers
to the stigmas of older ones. Flies also visit this
species to feed on the pollen, and though they cannot
obtain any nourishment from flowers in the latter
condition, still they sometimes come to them, apparently by mistake, and, must therefore occasionally
fertilise them. This species appears to have lost the
power of self-fertilisation.

I have already referred to *Lychnis vespertina* and
L. diurna in the first chapter. *L. Githago*, like
Dianthus, is adapted to butterflies. It agrees with
the flowers of that genus in the narrowness of the

tube, in the position of the honey, and in being distinctly proterandrous.

Silene nutans is a very interesting species. The life of the flower lasts three days, or rather, three nights. The first evening it opens towards dusk, becomes very fragrant, and expands its petals, while five of the ten anthers burst and expose their pollen. So it remains all night. Towards morning, however, the odour ceases, the petals shrivel and roll up, the stamens drop, and the flower looks dead. The next evening, however, it again opens, again emits a sweet scent, and the second series of five anthers open. Towards morning it again loses its smell, and again closes. The third evening it opens as before, but now the pistil has come to maturity, and the stigmas occupy the position, which the two previous nights had been filled by the anthers.

In *Silene inflata* (the Bladder Campion) there are, according to Axell (" Om Anord. för de Fan. Väx. Bef." p. 46), three kinds of flowers ; some with stamens only, some with a pistil only, some with both.

In *Stellaria graminea* (Fig. 62) the honey-glands are situated at the base of the five outer stamens. The flowers pass through three stages ; firstly , that in which the five outer stamens are mature, and incline towards the middle of the flower. In the second, the five inner stamens are mature. Lastly, the stigmas rise and expand themselves, while the stamens gradually shorten and shrivel up. Before this is accomplished, however, the stigmas have curled over and come into contact with the anthers, so that if the visits of insects are deferred, the flower fertilises itself. *Stellaria Holostea* is more conspicuous, and the three periods are more distinct, but the flower still retains the power of self-fertilisation.

In *S. media* (the Chickweed) the flowers are less conspicuous, and the five inner stamens are often rudimentary or entirely absent ; nay, two of the five outer ones are sometimes also rudimentary, though their honey-gland is always present. It is also proterandrous.

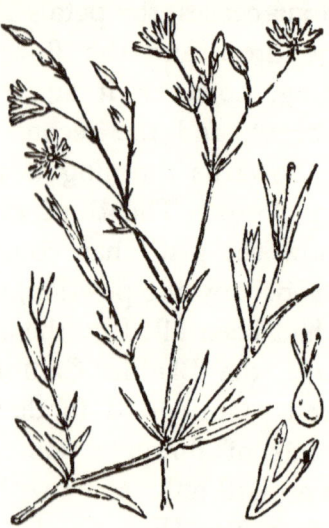

FIG. 62.—*Stellaria graminea.*

Cerastium arvense agrees with *Stellaria Holostea* in the position of the honey-glands, and in the sequence of development of the stamens and pistil. It is much frequented by insects. In other forms of this genus, as, for instance, in *C. semidecandrum* (which Bentham regards as a variety of *C. vulgatum*), the flowers are less conspicuous, and in consequence the visits of insects are fewer, the priority of the stamens is less marked, and self-fertilisation is more frequent.

The inner, honeyless stamens, which in *Stellaria minor* are often wanting, are in this form always rudimentary, according to Müller, while Bentham says that the whole number is often present. Both observers are so correct, that there is probably a difference in this respect between English and German specimens.

Sagina nodosa is also proterandrous ; while *Mæhringia trinervia* is, on the contrary, proterogynous.

The Caryophyllaceæ constitute therefore a very interesting and varied order. As a general rule the more conspicuous the flower, the more decided the dichogamy ; conversely, the smaller the flowers, and therefore the less frequent the visits of insects, the greater are the chances of self-fertilisation. The order also presents us with an interesting series commencing with open-flowered species, the honey of which is accessible even to beetles, and short-tongued flies, through those which are adapted to certain flies (Rhingia) and Bees ; to the species of *Dianthus, Saponaria*, and *Lychnis Githago*, the honey of which is accessible to Lepidoptera only.

HYPERICACEÆ.

There is only one British genus of this order, the well-known Hypericum, which, however, contains eleven British species. The stamens are united into bundles ; the styles are generally three in number, alternating with the bundles of stamens. In the large-flowered Hypericum, however (*H. calycinum*), the styles are five in number, and are raised above the stamens. *Hypericum perforatum* (the Common Hypericum) is so named from the remarkable peculiarity of having the leaves studded by pellucid dots; and several of the species have the sepals fringed with black or red glands. The flowers belonging to this genus are generally very conspicuous, both from their bright yellow colour and from their association in clusters. They secrete no honey, but are frequently visited by insects, partly for the sake of the pollen, partly perhaps in a vain search for honey. Under these circumstances, cross-fertilisation must frequently occur, though no doubt the flowers often fertilise themselves.

LINACEÆ.

This order contains two British genera, Linum, and Radiola; the former is the well-known flax, the latter a minute erect annual, which grows on heaths and sandy places. The genus Linum contains five British species, which differ considerably in the size of their flowers, from the beautiful, blue, common flax, to the minute *L. catharticum*, the petals of which are but little longer than the calyx, and which yet secretes honey, from five minute glands situated on the outer side, and near the base of the five stamens. It is therefore, in spite of its minute size, frequently visited by insects, though in their absence it is capable of self-impregnation. So far as has been hitherto observed *L. usitatissimum*, though differing so much from *L. catharticum* in the size of the flowers, agrees in general arrangement, and is also capable of self-fertilisation.

The crimson *L. grandiflorum*, on the contrary, as Mr. Darwin has shown (*Jour. Linn. Soc.*, Feb. 1863) presents two forms, which occur in about equal numbers, and differ little in structure, though greatly in function. In the one form, the column formed by the united styles and the short stigmas, is about half the length of the whole pistil in the other or "long-styled" form. The stigmas also of the short-styled form diverge greatly from each other, and pass out between the filaments of the stamens, thus lying within the tube of the corolla, while in the long-styled form the elongated stigmas stand nearly upright, and alternate with the anthers.

By a series of careful and elaborate experiments Mr. Darwin has shown that this species is almost entirely sterile with pollen of its own form. He repeatedly placed pollen of long-styled flowers on the stigmas of the same kind, and pollen of short-styled flowers on stigmas of short-styled flowers, but without effect; while if pollen of a long-styled flower is placed on a short-styled stigma, or *vice versâ*, abundance of seed is produced. In short, the pollen of the *L. grandiflorum* is differentiated, with respect to the stigmas of all the flowers of the same form, to a degree corresponding with that of distinct species of the same genus, or even of species of distinct genera.

Linum perenne is also dimorphous, and the difference between the two forms is more conspicuous.

MALVACEÆ.

Of this order we have three British genera, Lavatera, Althæa and Malva, with respectively one, two, and three specific forms. In the intoductory chapter, I have already called attention to the structure of the Mallow, with especial reference to the differences existing between *Malva sylvestris* (Figs. 43 and 45) and *M. rotundifolia* (Figs. 44 and 46). The honey glands are five in number, at the base of the flower. Althæa and Lavatera are said to agree in general structure with Malva.

TILIACEÆ.

Of this order we have in England only one species, the Common Lime (*Tilia Europæa*), which, however, is not a native species. The flowers are very sweet, and great favourites with bees. Their abundance and the size of the tree render colour unnecessary. The honey is secreted by the sepals, and is accessible even to short-lipped insects; while, as the flowers hang down, it is completely protected from rain. The stamens are numerous, but, as Hildebrand has pointed out, they have shed their pollen before the stigma is mature, and the flower is therefore incapable of self-fertilisation. The visits of insects are very numerous, and yet in this country the Lime seldom produces ripe seed.

This order contains four British genera ; Geranium, Erodium, Oxalis, and Impatiens.

The genus Geranium possesses a peculiar interest in the history of the present subject, because, as Sprengel tells us, the hairs in the corolla of *G. sylvaticum* (See p. 1), attracted his attention, and led to the researches which are so well described in his interesting work.

The flowers of the species of Geranium differ considerably in size : the larger flowered species, such as *G. sanguineum, G. phæum, G. pratense* (Figs. 40 and 42), and *G. sylvaticum*, being perennial, the smaller ones annual, or biennial. *Geranium palustre*, with which *G. pratense, G. sylvaticum*, and *G. sanguineum*, closely agree, is taken by Sprengel as a type of the large flowered species. The honey glands are five in number, situated near the base and at the outer side of the outer stamens; and are effectually protected by fringes of hairs arranged just above them, so as to prevent any rain from obtaining access to them.

The stamens are ten in number, of which one half are longer than the remainder : the pistil terminates in five lobes, the upper surfaces of which constitute the stigmas. The flower opens widely by day, hangs down, on the contrary, and partially closes at night. The petals are ornamented by purple lines, which serve as honey-guides, pointing to the honey glands. When the flower first opens (Fig. 41) the stigma is immature, and the five lobes are closely pressed together (*b*), so that the stigmatic surfaces are not exposed. Nor do they separate (Fig. 42), or become susceptible of fertilisation, until after the anthers have all shed their pollen. The flower, in fact, passes through three distinct stages : first, the five outer stamens open, and shed their pollen ; then the five inner ones; and lastly, after the pollen is all shed, the stigmatic surfaces expand and attain maturity. The flower therefore cannot fertilise itself.

On the other hand, in the smaller species of Geranium the stigmas come to maturity before the stamens have shed all their pollen ; hence the visits of insects are not so necessary, and hence, probably, the smaller size of their flowers. (See *antè* p. 48.)

There is also another difference, to which I will call attention. *G. Robertianum* does not possess the fringes of hairs by which the honey is in *G. sylvaticum* protected against the access of rain ; on the contrary, the petals are entirely glabrous. This difference is apparently connected with the form of the flower, which is less open than is the case with *G. sylvaticum* On the contrary, it forms a distinct tube, the entrance to which is sufficiently protected against rain by the stamens and pistil.

The smaller flowered species moreover offer remarkable differences among themselves. Thus *G. molle* and *G. pusillum* are at first sight very similar, and, as Bentham observes, are no doubt sometimes confused ; yet they differ remarkably. When *G. molle* first opens, the pistil is immature, and the stigmatic surfaces are closely appressed. The outer anthers then begin to open one after the other, so that the flower is for some time merely male. Before, however, the first five anthers have completely shed their pollen, the stigmatic surfaces arrive at maturity and expand ; so that, during the second period, the flower is both male and female.

In *G. pusillum*, on the contrary, when the flower first opens, the stigmatic surfaces are mature and expanded, but the anthers, are not yet ripe; the flower consequently is merely female, and can only be fertilised by pollen from an older flower. Soon,

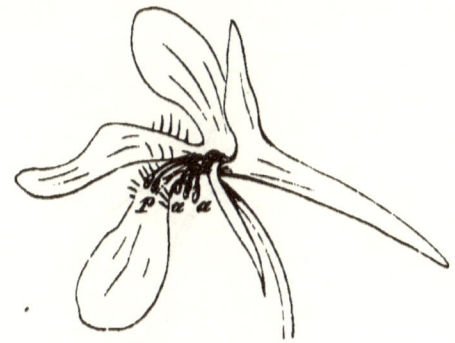

Fig. 63.—Young flower of *Tropæolum major*. Neither the anthers nor the stigma are yet mature.

however, the anthers of the five outer stamens ripen, and open, so that the flower is both male and female. Another remarkable difference is, that in *G. pusillum*, as in the allied genus Erodium, the five inner stamens produce no pollen.

In the genus Erodium, as in *G. pusillum*, the five inner stamens are rudimentary, and produce no pollen. The stamens ripen, however, before the stigma, though if the visits of insects be deferred, the flower is capable of self-fertilisation.

G

To this family also belongs Tropæolum, the common Nasturtium of our gardens. Here the honey is
contained in a long spur. The flower passes through
three well marked stages (Figs. 63–65). When it
first opens, as shown in Fig. 63, the anthers (*a*) are
unripe, the pistil (*p*) is short and immature. Soon,
however, one of the anthers ripens, opens, and
turns up, as shown in Fig. 64 *a, a,* so as to stand
directly in front of the opening to the tube; a
humble bee, therefore, or other insect of similar size,

Fig. 64.—Flower of *Tropæolum major* in the second stage. Some of the anthers
are now mature, and stand upright in front of the entrance to the spur.

visiting the flower for the sake of its honey, could
not fail to rub some of the pollen off on to her breast.
Shortly afterwards a second stamen ripens, and
assumes the same position, with the same result, and
the rest gradually follow. In flowers which I have
watched, this process occupies from three to seven
days, by which time the stamens have all come to
maturity, after which the anthers drop off, and the
filaments turn down as shown in Fig. 65, so as to be
well out of the way. It is now the turn of the pistil,

which in the meantime has elongated, and assumes the
position which the stamens had successively occupied ;
the result of which is, that a bee which had previously visited a younger flower and dusted her
breast with pollen could not fail to deposit some of
the pollen on the stigma.   It will be observed that
the lines on the flower as usual point to the honey.
The three lower petals bear a number of lanceolate
processes, which, as Sprengel has pointed out, serve

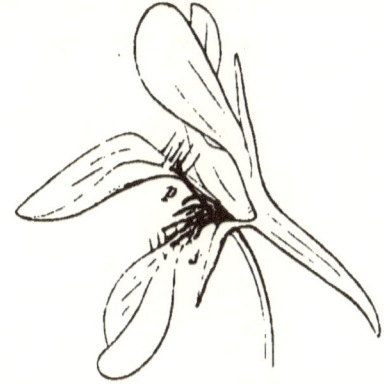

FIG. 65.— Flower of *Tropæolum major* in the third stage.   The anthers have all shed
their pollen, and the pistil now occupies the position in front of the entrance to
the spur.

to prevent rain from obtaining access to the tube,
and also perhaps in guiding the insects, so as to
bring their breasts truly against the anther or stigma.
The calyx, which from the position of the flower is
almost as much exposed as the corolla, is of the
same colour as the petals.

Oxalis acetosella is one of the species which produces
"cleistogamous" flowers (see *antè* p. 36).   This was
first, I believe, observed by Michelet (*Bull. Soc. Bot.
de France,* 1860, p. 465).

G 2

Hildebrand has shown that of the foreign species of Oxalis some are dimorphous, some trimorphous, This is, however, not the case with either of our English species, and I will therefore postpone any remarks on this curious arrangement until we come to some of the English species in which it occurs.

Impatiens noli me tangere is proterandrous, and the larger flowers cannot fertilise themselves. This species. however, also produces cleistogamous flowers. (Mohl. "Bot. Zeit.," 1863. Bennett, "Linn. Jour.," vol. xiii.) The seed capsules, when ripe, burst open if touched.

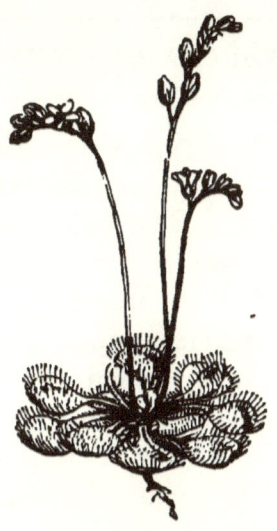

DROSERA ROTUNDIFOLIA.

CHAPTER IV.

CALYCIFLORÆ.

THIS subclass contains those Dicotyledons in which the perianth is double, the petals separate, and the stamens either perigynous or epigynous.

CELASTRACEÆ.

This order contains one British species, the Common Spindle Tree, *Euonymus Europæus*. The flowers secrete honey, and are frequented by Diptera (Flies) and Hymenoptera, especially the former. They are proterandrous.

RHAMNACEÆ.

The Vine and the Virginian Creeper belong to this order; but we have only one British genus, Rhamnus, (the Buckthorn) with two species, *Rhamnus cathar ticus* (the common Buckthorn and *Rhamnus frangula*), the Alder Buckthorn.   The two species differ considerably.   In *Rhamnus frangula*, the sepals, petals and stamens are five in number; the petals are very small.   The stamens open before the stigma is fully developed, and probably even before it is capable of fertilisation. The pistil is in the centre, and insects which visit the flower for the sake of the honey necessarily touch the stamens with one side of the proboscis and the pistil with the other. They must, therefore, often convey the pollen from one flower to another.   In the absence of insects, however, *R. frangula* is capable of self-fertilisation.

In *R. catharticus*, on the contrary, the flowers have four petals, and are diœcious; the male flowers have a rudimentary pistil, and the female flowers bear minute stamens.   The individual flowers are very small, they are rendered conspicuous by being associated in clusters, while those of *R. frangula* are in twos or threes.

R. lanceolatus, which, however, is not an English species, has been shown by Mr. Darwin to be diœcious (*Jour. Linn. Soc.*, v. vi., 1862, p. 95.)

LEGUMINOSÆ.

This is a very extensive order, containing eighteen British genera; the Peas, Vetches, Brooms, Clovers, Furze, &c. belong to it.

It is probable that all flowers which have an

irregular corolla are fertilised by insects.　The advantage of the irregularity is that it compels the insects
to visit the nectary in one particular manner.　In the
present group the result is that insects necessarily
alight on a particular part of the flower, when their
weight in many cases causes certain mechanical
effects by which the pollen is transferred to the body
of the insect, and thus carried from one flower to

Fig 66.—*Lotus corniculatus.*

another.　The corolla in the Leguminosæ consists of
five petals; an upper one, usually called the "Standard," two lateral ones, or "Wings;" and two lower
ones, united at their edges into a boat-shaped organ,
or "keel."

The bases of the stamens coalesce into a hollow
tube (Fig. 70 and 71 *t*) the inner walls of which, at
their base, secrete honey in some species, though not

in all. In the former, one or more of the stamens is detached, as in the Lotus (Fig. 70 *b*), or atrophied, so as

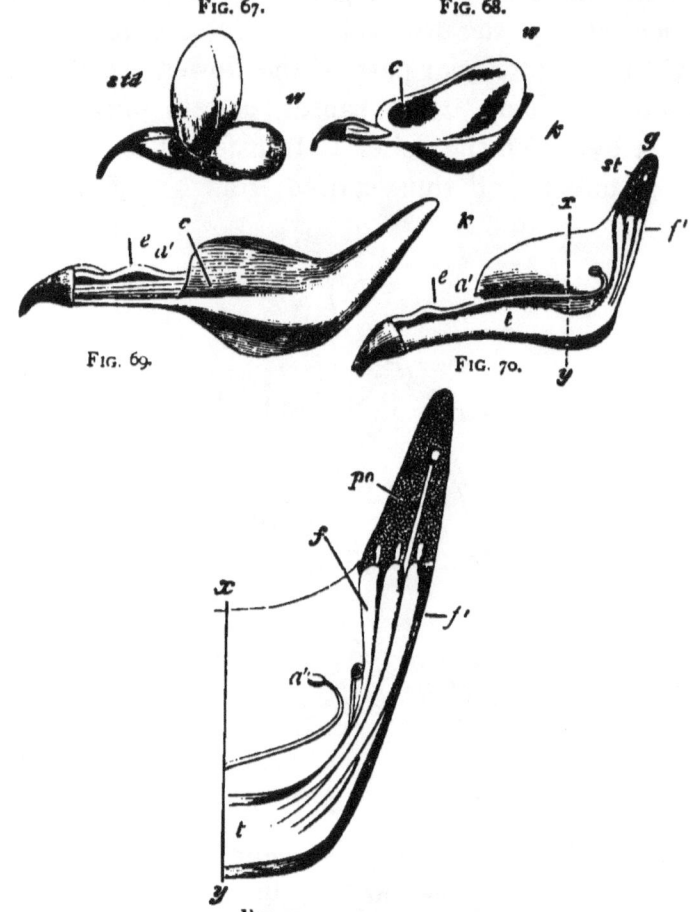

FIG. 67.　　　　　FIG. 68.

FIG. 69.　　　　　FIG. 70.

FIG. 71.

FIG. 67.—Flower of *Lotus corniculatus* seen from the side and in front.
FIG. 68.—Ditto, after removal of the standard.
FIG. 69.—Ditto, after removal of the standard and wings.
FIG. 70.—Ditto, after removal of one side of the keel.
FIG. 71.—Terminal portion of Fig. 70 more magnified.

　　e, entrance to the honey; *d*, the free stamen; *c*, the place where the wings lock with the keel; *f'*, expanded ends of stamens; *f*, filaments of stamens; *g*, tip of keel; *po*, pollen; *st*, stigma.

to leave a space through which bees can introduce

their proboscis into the tube.   In those species which
do not secrete honey this is unnecessary, and the
stamens are all fully developed and united.

In the Common Birdsfoot Trefoil (*Lotus corniculatus*) the anthers burst and emit their pollen before
the flower opens, and indeed before the petals have
attained their full size.   At this time the ten stamens
form two groups, five of them being longer than the
others ; but by the time the flower opens they are
all of the same length, though the five outer ones are
somewhat swollen at the end ; a difference which subsequently becomes still more marked.   The pointed
end of the keel is now filled by a mass of pollen
(Fig. 70 and 71 *po*), while the anthers, having discharged their contents, commence to shrivel up.   The
free ends of the five outer stamens continue, however,
to increase in size ; so that, with the pollen mass, they
completely fill up the cavity of the keel.   When the
flower opens the pistil, stamens, and pollen occupy
the position shown in Figs. 70 and 71.

The five inner stamens, having discharged their
pollen have become useless, shrivelled up, and lie in
the broader part of the keel ; the five outer ones, on
the contrary (Fig. 70 *e*), which still have an important
function to perform, lie behind the pollen mass, and
keep it in its place.

Insects do not generally alight directly on the keel,
but rather on the wings, which are more conveniently
situated ; the two, however, are relatively so arranged,
that when a bee alights on the wings, she presses
down the keel, which is locked with the two wings by
a projection and corresponding depression, as shown

in Figs. 68 and 69 *c ;* thus a portion of the pollen and also the tip of the pistil are forced out at the point of the keel, and against the breast of the bee, until on the removal of the pressure the elasticity of the flower causes the various organs to resume their former position ; an obvious advantage, which prevents the pollen from being wasted. The union of the stamens at their base has probably reference to this, as Sprengel has suggested. From the manner in which

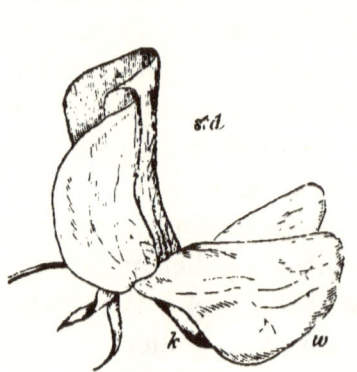

FIG. 72.—Flower of Sweet Pea, in its natural position.

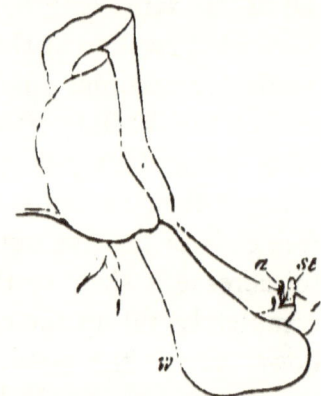

FIG. 73.—Ditto. The wings are depressed, the stamens and pistil exposed.

these flowers are evidently arranged with a view to the visits of insects, we can hardly doubt but that these visits are of importance to the plant.

In the Sweet Pea (Figs. 72 and 73), on account of its larger size the action is still more easily visible. Fig. 72 represents a flower in the natural position. Now if the two ends of the wings be taken between the finger and thumb, and pressed down, so as to imitate the effect produced by the pressure of an insect, the keel is depressed with the wings, while the pistil and stamens are thus partly uncovered, as shown in Fig. 73. When the pressure is removed, the flower resumes its former position.

Trifolium repens (the White Clover) agrees with Lotus in its general structure, but is somewhat simpler. The wings are actually united to the keel at one point. In *T. pratense* the flowers are longer, and the honey is only accessible to those bees which have a very long proboscis. As in other such cases, however, *Bombus terrestris* obtains access to it by eating a hole through the side of the flower. According to Darwin this species is only fertilised by humble bees, but Delpino disputes this. *Trifolium subterraneum* has small cleistogamous flowers (*Mohl. Bot. Zeit.*, 1863) besides the usual ones.

Anthyllis vulneraria also agrees with Lotus in its general arrangement. The tube of the flower is, however, elongated ; and in consequence, this species is only visited by bees with long tongues. In the young flower, though the pistil is in the keel, and necessarily in contact with the pollen, H. Müller has observed that the stigma is dry, and that no pollen adheres to it. Subsequently, however, when most, or all, of the pollen has been removed, the stigma becomes sticky, and pollen adheres closely to it.

In Ononis (the Restharrow) the general arrangement is very similar. There are, however, several important differences. Ononis does not secrete honey, and consequently there is no need for the separation of the upper stamen, which in this genus is attached to the rest. Again, in Ononis all the stamens are thickened at the end ; the outer ones, however, much more so than the inner ones. The inner ones, on the contrary, produce much more pollen than the others; a difference of function which is even more marked in the Lupins.

Ononis is exclusively fertilised by bees, and H. Müller has repeatedly seen male bees visiting this species in a vain search for honey.

Onobrychis sativa (the Common Sainfoin) agrees with *Trifolium repens* (Clover) in its general structure; but the wings are greatly reduced in size and appear to serve only in preventing the honey being reached from the side, or at least in rendering this more difficult. This species is sufficiently conspicuous, and as the honey is accessible even to insects with a short proboscis, it is much visited. When mature, the stigma projects 1 to $1\frac{1}{2}$ m. beyond the keel, and according to H. Müller the flower has lost the power of self-fertilisation.

In *Genista tinctoria* the ten anthers lie in two distinct rows. While the flower is still in the bud, the four upper anthers of the outer row are already on the point of opening, while those of the inner circle have not nearly reached their full size. These four anthers now open and shed their pollen into the space at the apex of the keel, after which they shrivel up. The fifth, although it has attained its full size, remains closed. The next process is that this anther and those of the second row also open, and the pollen occupies the end of the keel between the anthers and the stigma, as in Lotus. While, however, in Lotus when the insect leaves the flower and the pressure is thus removed, the keel resumes its position,

and the stamens and pistil are again protected; in
Genista tinctoria, on the contrary, the flower opens
once for all. The keel is at first nearly parallel to
the standard (Fig. 74). This position is, however,
one of tension; the keel is retained in it by the union
of its upper margins, which inclose and retain the
curved pistil which presses against them like a spring.
The sides of the keel have near the base a projecting
lobe (Fig. 76 *m*), which locks with one at the corresponding part of the wing. When an insect, alighting
on the flower presses open the keel in search of pollen,
as soon as the curved end of the pistil is set free, it

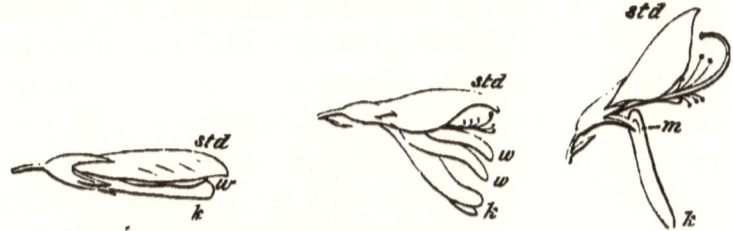

FIG. 74.—Flower of *Genista tinctoria* unopened.
FIGS. 75, 76.—Ditto, opened. *std*, standard; *w*, wing; *k*, keel; *m*, projection on keel.

springs up with a jerk, the keel, on the contrary,
springs back (Figs. 75 and 76), and the pollen is ejected
in a shower. It appears that the flowers do not open
of themselves if insects are prevented from visiting them (Henslow, *Jour. Linn. Soc.*, v. x. p. 468).

Genista tinctoria contains no honey, and yet it is
visited by several insects which do not consume pollen.

The flowers of the Common Furze (*Ulex Europæus*)
agree in essentials with those of the preceding species.
The calyx, however, is larger, and coloured; the
wings are longer in proportion and project beyond the

keel. They also lock at the base with the keel, and
when they are pressed downwards the flower bursts
open. The Furze has, like Cytisus and the Broom,
on the outer part of the staminal lobe a honey-
containing tissue.

In the Laburnum, the tip of the pistil is protected
from its own pollen by a ring of close hairs; when,
however, the pistil has grown to the very point of
the keel, these hairs shrivel and turn outwards, so as
to expose the tip of the pistil, which thus comes in
direct contact with the breast of any bee which may
alight on the flower.

In the Broom (*Sarothamnus scoparius*) the flowers
also explode. If, however, the bee alights on a newly-
opened blossom, the shorter stamens only emerge
and dust the abdomen of the insect. If, on the contrary, the flower is a day or two older, the pistil and
longer stamens spring out, and the hairs on the pistil
deposit pollen on the bee's back. The pistil gradually
curls, and the stigmatic surface turns up, so as to
stand close to the anthers of the shorter stamens.
In this position it is so placed that it would come in
contact with the abdomen of the bee. "Thus," says
Mr. Darwin, "both the upper and lower surface of
the bee get dusted with pollen, which will be transferred to the stigma at two different periods." (*Linn.
Jour.* v. ix. p. 358.)

In *Medicago sativa*, as in Genista and the Broom,
the flowers open once for all; but the elastic power
is confined to the upper stamens. In the Broom and
in Genista, the resistance is obtained by the union of
the upper edges of the keel. These are also united
in Medicago, but even if they are separated no ex-

plosion takes place; the flower being still locked together by four processes, two of which point forwards and two backwards. These fit so beautifully that the proboscis of a humble bee is sufficient to unlock them, and release the stamens; though, according to Henslow, (*Linn. Jour.* 1866, p. 328) the hive bee is unable to do so. Hildebrand, however, has observed that in the absence of insects, it fertilises itself. In *M. lupulina* the elasticity is much less than in *M. sativa*. Medicago is a honey-bearing genus.

In the Leguminosæ hitherto mentioned, when the keel is forced open, both stamens and pistil emerge from it. In Lathyrus (the Pea), however, this is not the case. In *L. pratensis*, for instance, the stamens do not leave the keel, but the pistil is provided with a brush of hairs, which sweep the pollen before them.

In the Scarlet Runner (*Phaseolus communis*) which has been described by Farrer (*Ann. and Mag. of Nat. His.* 1868, p. 255), the keel is spiral, as well as the stamens and pistil. The former are weak, and never protrude; while the pistil, on the contrary, is stout, strong, and very elastic. In the natural position, the stigma just protrudes out of the mouth of the keel, while the terminal portion of the style within the tube is covered with fine hairs. When, therefore, the bee alights on a flower, and inserts her proboscis into it, the stigma will come in contact with the base of the proboscis, and will sweep off any pollen which may be adhering to it. As, however, the bee presses more on the flower, in its efforts to get the honey, the pistil comes further out of the flower; the stigma turns upwards, away from the insect, and the brush of hairs, which has swept the sticky honey out of the anthers, and is consequently covered with it, rubs against the head of the bee and the base of the proboscis, on which it deposits a certain quantity of the pollen, to be again transferred to the stigma of the next flower which the bee visits.

The Common Pea (*Pisum sativum*) is said not to be well adapted to our British bees. Its structure, probably, has reference to some of the larger southern species.

In *Vicia cracca* each wing is united to the keel in two places. Though the parts of the flower fit closely to one another, still from the smallness of its size the honey is accessible to most bees ; and, owing to the conspicuousness of its bunches, it is much visited by them. From their arrangement and elasticity, the various parts of the flower resume their original position after each visit.

Vicia sepium, in general characters, agrees with *V. cracca*, though the arrangement of the hairs on the pistil is very different. The insects by which it is visited are, however, much fewer. Its larger size, coupled with other minor differences, excludes flies, Lepidoptera, and the smaller bees. Even *Bombus terrestris* (the Common Humble Bee) does not attempt to suck it, but bites a hole through the side. In *V. faba* the wings and keel are less closely united, and the honey is more easily accessible. The flower also is less elastic, and if opened widely does not again resume its original form.

It appears then that the Leguminosæ are all adapted to fertilisation by bees, and, as Delpino has pointed out, the flowers fall into four series.

1. Those in which the pressure of the bee pumps out, as it were, a certain quantity of pollen; the flower resuming its original form when the pressure is removed. (Lotus, Anthyllis, Ononis, and Lupinus.)

2. Those in which not only the pollen, but also some of the stamens are pressed out; the flower resuming its form on the removal of the pressure, as in the first division. (Melilotus, Trifolium, Onobrychis.)

3. Those in which the flower bursts on pressure and ejects the pollen. (Medicago, Genista, Sarothamnus.)

4. Those in which, on the pressure of the bee, the pollen is swept out by a brush of hairs situated on the pistil. (Lathyrus, Vicia, Pisum, Phaseolus.)

The power of self-fertilisation seems to be lost in some species of Phaseolus, Onobrychis, and Saro-thamnus; and to be much diminished in others, as in *Trifolium repens* and *Vicia faba.*

ROSACEÆ.

This order contains seventeen British genera, includ-ing Prunus (the Cherry, &c.), Spiræa, Geum, Rubus (Blackberry, &c.), Fragaria (Strawberry), Potentilla, Alchemilla, Sanguisorba, Poterium, Agrimonia, Rosa, Pyrus, Cratægus, &c.

Prunus. Our three species of this genus differ somewhat in the relations of the anthers to the stigma. In *P. cerasus* (the Cherry) both mature at the same time, while in *P. spinosa* (the Black-thorn) and *P. padus* (the Bird Cherry) the stigma reaches maturity before the anthers: though as it retains the capability of fertilisation after the anthers have opened, the flowers are doubtless often self-fertilised; which from the position of the anthers probably happens more frequently in the Bird Cherry than in the Black-thorn. The flowers are melliferous. The British species of Spiræa, on the contrary, contain no honey, but are rich in pollen and are consequently visited by insects; which, from the weakness of the petals, generally alight on the stigma, and thus effect cross-fertilisation; though the flowers, if not visited by insects, fertilise themselves. Among the foreign species of this genus, several are melliferous.

Both our English species of Geum (*G. rivale* and *G. urbanum*) are melliferous: but the flowers of *G. rivale* are much larger than those of *G. urbanum*, and more frequently visited by insects. Müller mentions that *Primula elatior* is deserted by bees as soon as *Geum rivale* comes into flower. Van Tieghem states that while *G. urbanum* produces honey in the north, this is not the case in France, at least, near Paris.

The genus Rubus is very variable, and there are great differences of opinion among botanists as to the specific limits, and the number of species. Bentham admits five, though even these, he adds, "will very frequently be found to pass imperceptibly one into the other." The Raspberry (*Rubus idæus*) is so called because it is said to be very frequent on Mount Ida, where in 1872 Mr. (now Sir M. E.) Grant Duff and I found in abundance a species, which if not identical with, was very near, our *R. idæus*. This species, though it secretes honey, is not apparently a great favourite with insects, and frequently fertilises itself. The flowers of the Blackberry (*R. fruticosus*), on the contrary, are much more conspicuous, and the stamens are turned more out-wards, so as to leave more room between themselves and the pistil. They are very much frequented by insects, and as the stamens ripen gradually, and from the outside inwards, there is a considerable interval during which, though the pistil is mature, and some of the anthens are ripe, self-fertilisation is difficult ; while from the great frequency of insect visits, fertilisation is generally effected before the inner anthers are mature.

In the Strawberry (*Fragaria vesca*) also, the stigma arrives at maturity some time before the anthers, so that cross-fertilisation generally takes place. The species of Potentilla agree with Fragaria in habit, foliage, and flowers, but the fruit is not succulent. The honey, however, is not secreted in drops, but forms a thin layer. According to Van Tieghem, *P. tormentilla* produces honey abundantly in the north, but scarcely any in the neighbourhood of Paris. *Agrimonia Eupatoria* appears to secrete no honey, and is but seldom visited by insects. *Alchemilla vulgaris* is remarkable for variability. The honey is scanty, so that it is little visited by long-lipped insects ; while, from its greenish colour, it is not attractive to beetles, or other colour-loving species. Self-fertilisation is, however, comparatively rare, since the flowers seldom possess both anthers and stigmas ; one or the other being generally more or less rudimentary. This plant, therefore, may be considered to be becoming dioecious.

The next two genera of Rosaceæ, Sanguisorba and Poterium, each of which contains a single British species, have been already alluded to in the opening chapter (*ante*, p. 10). *Sanguisorba* (Fig. 10) *officinalis* is monœcious and fertilised by insects. In *Poterium sanguisorba* (Fig. 9) some flowers are male, some female, and some hermaphrodite, and the pollen is said to be wind-borne. In other respects these two plants are curiously similar.

There is almost as much difference of opinion with reference to the specific limits in the genus Rosa as is the case in Rubus. Bentham admits five British species, while others, as, for instance, Babington, extend the number to fifteen or twenty. The flowers do not appear to secrete honey, but are much visited by insects for the sake of the pollen. The numerous stamens ripen at the same time as the pistil, but from the convenient position of the latter, insects very frequently alight upon it, and thus fertilise it with pollen from other flowers, though self-fertilisation probably often occurs.

Pyrus malus (the Apple), on the contrary, and *Cratægus oxyacantha* (the Hawthorn) are melliferous, and the stigma comes to maturity before the anthers.

ONAGRACEÆ.

This order contains six English genera, Epilobium Œnothera, Ludwigia, Circæa, Myriophyllum, and Hippuris.

The instructive differences which exist between the different species of Epilobium have already been referred to in the introductory chapter. *Œnothera biennis* is really a North American plant, though now naturalized in some parts of England. As its name denotes (Evening Primrose) it is a yellow night flower; it secretes honey, and is probably fertilised by moths, though it remains open by day, and is also visited by bees. Ludwigia contains a single species,

L. palustris—a minute marsh plant, hitherto found in very few localities in Britain, though it ranges over Central Europe, Asia, and North America. The genus Circæa contains two species, *C. alpina* and *C. lutetiana*, the Enchanter's Nightshade. This species has two stamens, and as the flower is small, any insect of moderate size would probably touch both them and the pistil ; most likely, however, coming in contact with the stigma first, as it projects rather beyond the anthers.

LYTHRARIEÆ.

This order contains two British genera, Lythrum and Peplis, the former of which is of peculiar interest and has been already alluded to in the opening chapter (*antè* p. 40).

Lythrum salicaria (Fig. 77), presents us with three distinct forms of flower, which were already recorded by Vaucher, while their functions and relations were first explained by Mr. Darwin. He distinguished them according to the length of their styles, as the Long-styled (Fig. 78), Mid-styled (Fig. 79), and Short-styled (Fig. 80). In this species it is remarkable that the seeds of the three forms differ from one another; 100 of the long-styled seeds being equal to 121 mid-styled, or 142 short-styled. The pollen grains, also, not only differ in size, the long stamens having the largest pollen grains, the middle-sized stamens middle-sized pollen grains, and the short stamens small pollen grains; but also in colour, being green in the longer stamens, and yellow in the shorter ones; while the filaments are pink in the long stamens, uncoloured in the shorter ones.

Mr. Darwin has also proved by experiment that this species does not set its seeds, if the visits of insects are prevented; in a state of nature, however, the plant is much frequented by bees, humble bees,

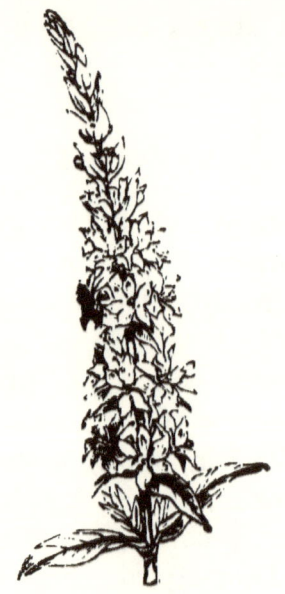

FIG. 77.—*Lythrum salicaria.*

and flies; which always alight on the upper side of the flowers on the stamens and pistil. Mr. Darwin has shown that perfect fertility can only be obtained by fertilising each form with pollen from pistils of the corresponding length.

Thus the long-styled form is naturally fertilised by pollen from the long stamens of the two other forms;

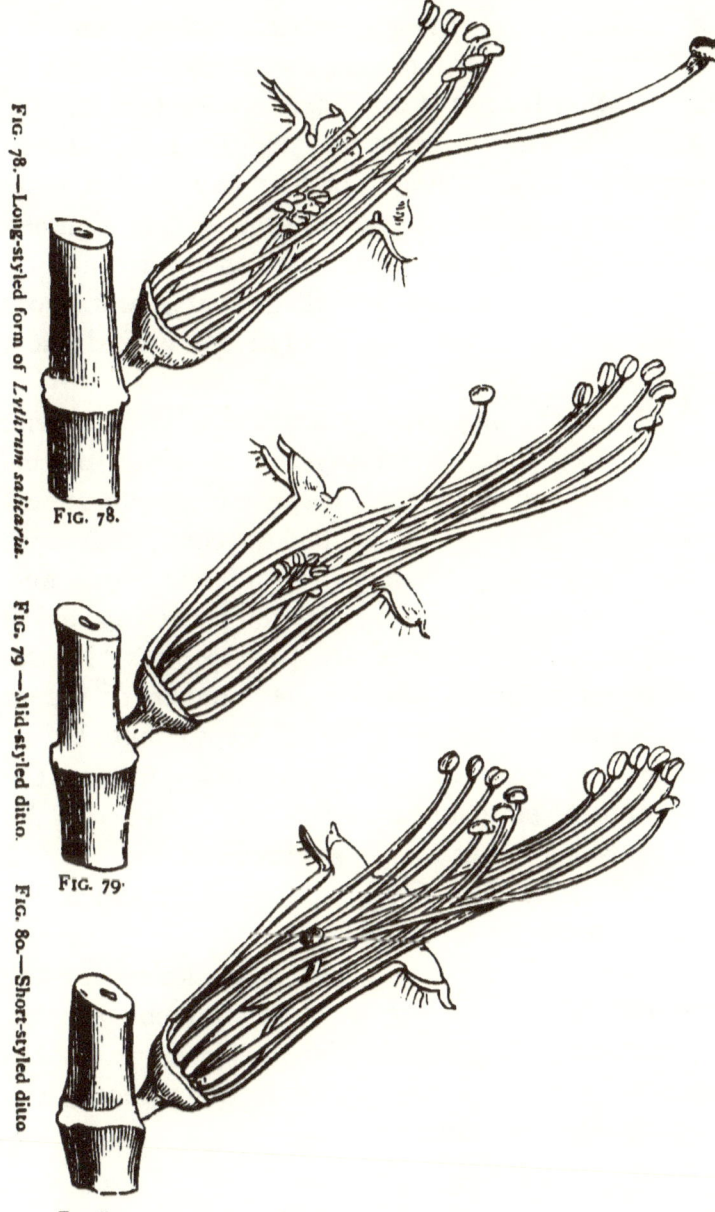

FIG. 78.—Long-styled form of *Lythrum salicaria.*

FIG. 78.

FIG. 79.—Mid-styled ditto.

FIG. 79.

FIG. 80.—Short-styled ditto.

FIG. 80.

but it can be so, though imperfectly, by its own two sets of stamens, and by the shorter stamens of the two other forms; it can, therefore, be fertilised, to use Mr. Darwin's expression, "legitimately" in two ways, and "illegitimately" in four ways. The same is the case with the other two forms, so that eighteen modes of union are possible, of which six are natural or "legitimate," twelve are illegitimate, and more or less sterile. This case is therefore indeed most complex.

Mr. Darwin suggests (*Jour. Linn. Soc.* v. viii. 1864, p. 193) that the *trimorphous* condition of this plant may be advantageous, because if it were dimorphous only there would be but an equal chance in favour of any two plants being of different forms, and therefore capable of self-fertilisation; whereas, being trimorphous, the chances are two to one. In the cowslip and primrose, where large numbers of plants grow together, this, he thinks, would not be so material. However this may be, the stigma and the two groups of stamens appear to correspond with the three divisions of the body (viz. the head, thorax, and abdomen) of the bee, *Cilissa melanura,* by which it is almost exclusively fertilised.

The genus Lythrum is also remarkable for the great differences existing between different species. For instance, *L. græfferi,* like *L. salicaria,* is trimorphous; while *L. thymifolia* is dimorphous, and *L. hyssophifolia* is homomorphous.

CUCURBITACEÆ.

Of this order we have only a single species, the common Bryony (*Bryonia dioica*). The flowers are diœcious, the males in small clusters, pale yellow, about half an inch in diameter; the females much smaller. Both secrete honey.

CRASSULACEÆ.

Of this order there are four British genera : Tillæa, Cotyledon, Sedum, and Sempervivum. The first two contain a single species each. Of Sedum we have nine species. Though the flowers are small, yet from the localities they occupy, and from their bright colours they are somewhat conspicuous, and are visited by many insects for the sake of their honey, which is accessible even to those with short tongues. Some (*S. acre, reflexum*, and *telephium*) are proterandrous, while *S. atratum*, according to Ricca, is proterogynous ; and *S. rhodiola* is diœcious.

RIBESIACEÆ.

This order consists, as far as Britain is concerned, of the genus Ribes, containing four species, the Gooseberry (*R. grossulariata*), Red Currant (*R. rubrum*), Black Currant (*R. nigrum*), and Mountain Currant (*R. alpinum*). They all supply honey. *R. grossulariata* is proterandrous, and is said to have lost the power of self-fertilisation. In *R. rubrum* and *R. nigrum* the stamens and pistil come to maturity simultaneously. *R. alpinum*, on the contrary, is diœcious ; and it is interesting that, according to Müller, this species is more frequented by insects than any of the others.

SAXIFRAGACEÆ.

An extensive order, ranging nearly over the whole world, but represented in Britain by only four genera, Saxifraga, Parnassia, Drosera, and Chrysosplenium.

The species of the genus Saxifraga are melliferous, and proterandrous. Bergenia (Saxifraga) crassifolia, which, however, is not British, though frequently grown in gardens, is according to Engler, proterogynous. In Chrysosplenium the anthers and stigma ripen simultaneously. *Parnassia palustris*, as its name indicates, inhabits wet and boggy places. It has ten stamens, of which however five only bear anthers, while the others secrete honey at the base, and terminate in from eight to seventeen beautiful yellow globular glands. These glands so closely resemble drops of honey that it is difficult to believe they are perfectly dry. They probably serve as sham drops of honey to attract flies. The five polliniferous anthers ripen, not simultaneously, but successively, and "as each ripens it places itself right on the

FIG. 81 —*Drosera rotundifolia.*

top of the stigma, with its back to it, and the pollen is then discharged from the anther on the side away from the stigma, so that it is scarcely possible for any to fall on it; and this is done by each of the five stamens in succession " (Bennett, " How Flowers are Fertilised," 1873, p. 19). The flowers are much visited by insects, especially by flies.

In the cases we have hitherto considered, the relation between the flowers and insects is one of

mutual advantage. The honey of the flowers affords
to the insects a rich and nutritious food; and if the
latter rob the flowers of some of their pollen, they
make ample amends by carrying a portion of the
remainder from one flower to another, and thus con-
ferring on the plant the great advantage of cross-
fertilisation. In Drosera (Fig. 81), on the contrary,
we find a very different state of things, for the plant

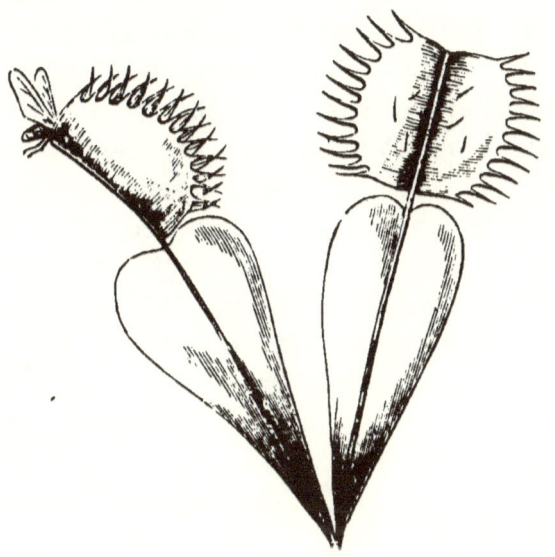

FIG. 82.—Two leaves of Dionæa : one open, one closed upon a fly.

catches and devours insects. This genus, and the
other plants which have this remarkable habit, have
recently been the subject of an admirable memoir,
by Dr. Hooker, read before the British Association (*Nature*, Sep. 3, 1874). The first observation
on insect-eating flowers was made, about the year
1768, by our countryman Ellis, on Dionæa, a North
American plant, the leaves of which have a joint

in the middle, and thus close over (Fig. 82), kill, and actually digest any insect which may alight on them. The plant has recently been studied by an American botanist, Mr. Canby, and, says Dr. Hooker, "by feeding the leaves with small pieces of beef, he found, that these were completely dissolved and absorbed ; the leaf opening again with a dry surface, and ready for another meal, though with an appetite somewhat jaded. He found that cheese disagrees horribly with the leaves, turning them black, and finally killing them. Finally, he details the useless struggles of a curculio (beetle) to escape, as establishing the fact that the fluid is secreted, and not the result of the decomposition of the substance which the leaf has seized. The curculio being of a resolute nature, attempted to eat his way out—'when discovered he was still alive, and had made a small hole through the side of the leaf, but was evidently becoming very weak. On opening the leaf, the fluid was found in considerable quantity around him, and was without doubt gradually over-coming him. The leaf being again allowed to close upon him, he soon died.'" Prof. Burdon Sanderson has recently made some interesting observations on the electrical changes by which these movements are accompanied. (*Brit. Ass. Report*, 1873.)

In the genus Drosera (Fig. 81), the hairs which cover the leaf, fold over and capture insects. This was first observed almost simultaneously by Mr. Whately and Mr. Roth. The latter says, "I placed an ant upon the middle of the leaf of *D. rotundifolia*, but not so as to disturb the plant. The ant endeavoured to escape, but was held fast by the clammy juice at the points of the hairs, which was drawn out by its feet into fine threads. In some minutes the short hairs on the disc of the leaf began to bend, then the long hairs, and laid themselves upon the insect. After a while the leaf began to bend, and in some hours the end of the leaf was so bent inwards as to touch the base. The ant died in fifteen minutes, which was before all the hairs had been bent themselves." Mr. Darwin has recently shown that while the leaves will in this way close over, and actually digest pieces of meat or other animal matter, they take little notice of inorganic substances.

I cannot pass from this subject without mentioning another insectivorous plant, the genus Sarracenia, though it is not British, and does not belong to the present order. *S. variolaris* has some of the leaves in the form of a pitcher which secretes a fluid, and is lined internally with hairs pointing downwards. Ants, flies and other insects which fall into this pitcher cannot get out again, and are actually digested by the plant. Up the outside of the pitcher there is a line of honey glands, which lure the insects to their destruction. Bees, however, appear to be scarcely ever caught.

UMBELLIFERÆ.

This is a very extensive order, containing no less than thirty-seven British genera (Carrot, Chervil, Parsley, Parsnip, &c.) and a very large number of species. The plants belonging to this group possess two great advantages—namely, firstly, the association of the numerous small flowers into comparatively large flat heads, by which they are made much more conspicuous : and, secondly, they all secrete honey in the centre of the flower on a flat disk (Fig. 84, 85) which is thus accessible to all insects, even those with the shortest lips. This is an advantage, as it effects a considerable saving of time, enabling the insects to visit a given number of flowers

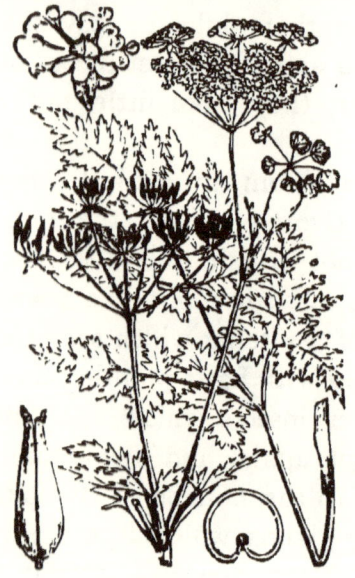

Fig. 83.—Wild Chervil (*Chærophyllum sylvestre*).

more rapidly, and consequently rendering their fert-ilisation more certain than if they had stood singly. But though the order is so rich in genera and species, it is comparatively uniform, and the divisions are for the most part characterised by the form and structure of the fruit. The flowers are generally small; the petals five, inserted round a little fleshy disk ; the stamens, also five, alternating with the petals.

The self-fertilisation which, in small flowers such as these, would otherwise naturally occur, is provided against by the fact that the flowers are generally pro-terandrous, the stamens ripening before the pistil, and the latter not being mature until the former have shed their pollen ; as, for instance, is shown in the following enlarged figures of the Wild Chervil (*Chæro-phyllum sylvestre*). Fig. 84 represents a floret in the earlier (male) condition, showing three ripe (*a′*) and

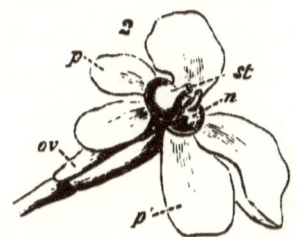

Fig. 84.—Flower of the Wild Chervil in the first (male) state.

Fig. 85.—Ditto, in the second (female) state.

two still immature (*a*), while the stigmas have not yet made their appearance : in Fig. 85 is represented the same flower in a more advanced condition, the stamens having fallen off, and the stigmas (*st*) being now mature. In some cases, flowers in both conditions may be found in the same head or umbel ; in others, as, for instance, in Myrrhis, the flowers of one head are all firstly in the male condition, and subsequently in that with mature stigmas, none of them arriving at the second stage until they have all passed through the first.

It will be seen that in these florets the petals are

not symmetrical, the outer ones being considerably larger than the others, and in many Umbellifers the florets themselves, on the outer edge of the umbel, are considerably larger than the inner ones. This distinction is carried still further in the Compositæ, where also the florets are so closely packed together that the whole flowerhead is commonly, though of course incorrectly, spoken of as a flower.

H. Müller has recorded 73 species of insects as frequenting the Wild Chervil. In some cases the number was even greater, as for instance in Heracleum, on which he has observed no less than 118. That the number depends on the conspicuousness of the umbel he illustrates by the following series, arranged in the order of the conspicuousness of the flowers,—viz., 1. Heracleum, 2. Ægopodium, 3. Anthriscus (Chærophyllum) sylvestris, 4. Daucus, 5. Carum, 6. Chærophyllum temulum, 7. Torilis. On these he found the following number of species of insects :

Heracleum	118
Ægopodium	104
Anthriscus sylvestris	73
Daucus	61
Carum	55
Chærophyllum temulum	23
Torilis	9

The position of the honey on a flat disk, which renders it accessible to most insects, has the opposite result as regards the Lepidoptera, which therefore, as might naturally be expected, are but rare visitors of the Umbelliferæ. I have sometimes wondered whether the neutral tints of these flowers have any connection with the number of species by which they are frequented.

TABULAR VIEW OF THE INSECTS VISITING SOME OF THE COMMONEST SPECIES OF COMPOSITES AND UMBELLIFERS.

	1	2	3	4	5	6	7	8	9
						Percentage of species belonging to			
	Whole number of species observed to visit the flowers.	No. of Lepidoptera (Butterflies and Moths).	No. of Apidæ (Bees).	No. of Diptera (Flies).	No. of Insects belonging to other groups.	Lepidoptera (Butterflies and Moths).	Apidæ (Bees).	Diptera (Flies).	Other Insects.
COMPOSITÆ.									
Taraxacum officinale .	93	7	58	21	7	7.5	62.5	22.6	7.4
Cirsium arvense . . .	88	7	32	24	25	7.9	36.4	27 3	28.4
Achillea millifolium . .	87	6	30	21	30	6.9	34.5	24.1	34 5
Chrysanthem. leucanth.	72	5	12	28	27	6.9	16.6	38 9	37.5
Centaurea jacea . . .	48	13	28	6	1	27	58.7	12 5	2
Carduus acanthoides .	44	4	32	3	5	9.1	72.7	6.8	11.3
Senecio jacobæa . . .	40	3	16	18	3	7.5	40	45	7.5
Picris hieracioides . .	29	3	16	9	1	10.3	55 2	31	3.4
Tanacetum vulgare . .	27	5	7	7	8	18.5	25.9	25.9	29.6
Eupatorium cannabinum	18	9	2	6	1	50	11.1	33.3	5.5
UMBELLIFERÆ.									
Heracleum sphondylium	118	0	13	49	56	0	11	41.5	47.4
Ægopodium podograria	104	0	15	34	55	0	14.4	32.6	52.9
Anthriscus sylvestris .	73	0	5	26	42	0	6.8	35.6	57.5
Daucus carota . . .	61	2	8	19	32	3.3	13.1	31.1	52.5
Carum carvi	55	1	9	21	24	1.8	16 4	38 2	43.6
Anethum graveolens .	46	0	6	15	25	0	13	32.6	54.3
Sium latifolium . . .	32	0	0	20	12	0	0	62.5	37.5
Angelica sylvestris . .	30	1	2	11	16	3.3	6.6	36.6	53.3
Chærophyllum temulum	23	0	1	10	12	0	4.3	43.5	52.2
Pimpinella saxifraga .	23	0	3	8	12	0	13	34.8	52.2

No order of plants are more visited by insects than the Compositæ and the Umbelliferæ; but from the difference in the form of the flowers the species are

I

very different. In the Umbellifers the honey, being secreted on an open disk, is therefore open to all insects. Though the tubes of the florets of the Compositæ are short, still the honey is not quite so accessible as in the Umbellifers. H. Müller gives the preceding table, which brings this out very clearly, and which also shows the care and perseverance with which he carried on his observations.

Thus, then, while in Centaurea, out of every 100 insects by which the flower is visited, no less than 58 are bees, 27 are butterflies or moths, 12 are flies, and only 2 belong to other groups; in the common Carrot on the contrary, where the honey is quite exposed, 13 in a hundred only are bees, 3 are butterflies or moths, 31 are flies, and 52 belong to other orders. If a flower with a longer tube than that of Centaurea had been selected for comparison, the difference would have been even more striking.

ARALIACEÆ.

The only European species belonging to this order is the Common Ivy (*Hedera helix*). It is proterandrous, and is much visited by flies and wasps.

CORNACEÆ.

This order contains one British genus, Cornus, with two species, *C. suecica* the Dwarf Cornel, and *C. sanguinea* the Common Cornel. The two species are very unlike; *C. suecica* being a low herb with minute flowers, which, however, are surrounded by four large, white bracts, which look like petals, and thus give the whole umbel the appearance of a single flower. *C. sanguinea* is a shrub which attains a height of five or six feet. The honey is secreted from a fleshy ring at the base of the pistil; it is accessible to all insects, and is much more visited by flies than by bees. The anthers and stigma mature simultaneously.

LAMIUM ALBUM.

CHAPTER V.

COROLLIFLORÆ.

This subclass contains those dicotyledons in which the petals are united together, at least at the base.

CAPRIFOLIACEÆ.

THIS order, which contains five British genera, Adoxa, Sambucus, Viburnum, Lonicera, and Linnæa, offers remarkable differences, especially in relation to the honey glands. Adoxa is a low, glabrous, light green herb : the flowers, which are coloured like the rest of the plant, secrete honey, which is exposed, and accessible to all insects. *Sambucus nigra* (The Common Elder), on the contrary, secretes no honey. It is nevertheless sweet-scented, and is visited by several insects, but often fertilises itself, as the stamens and pistil ripen simultaneously. Viburnum (the Guelder Rose) secretes honey, and the flowers are collected into a head as in the Elder, but the outer florets have the corolla considerably enlarged at the expense of the stamens and pistil. Although, therefore, they produce neither pollen nor seeds, they are useful to the plant, by rendering the other flowers more conspicuous, and thus attracting insects. In remarkable contrast to these species, with their exposed honey, is the genus Lonicera (the honeysuckle). *Lonicera caprifolium* has a honey tube no less than 30 mm. long, for the most part not above 1—2 mm. wide, and moreover a great part occupied by the style. It is often, however, half full of honey. As in the longest tongued bees (*Bombus hortorum* and *Anthophora pilipes*), the proboscis only attains a length of 21 mm., those of Flies (*Rhingia, Bombylius discolor*) not more than 11—12 mm., they are none of them in a position to extract all the honey; and in fact Müller never found them attempting to do so, though they visit the flowers for the pollen. The honey of *Lonicera caprifolium* is therefore especially adapted for the larger moths. The flowers open in the evening, and are then specially fragrant. Müller found the following moths on this species: *Sphinx convolvuli; S. ligustri; S. pinastri; Deilephila elpenor; D. porcellus; Smerinthus tiliæ; Dianthœcia capsincola, Cucullia umbratica, Plusia gamma, Dasyclura pudibunda.*

L. periclymenum (the Common Honeysuckle) agrees in most respects with the preceding species,

but the tube is rather shorter, and the honey in consequence more accessible to bees. In our third species
again, *L. xylosteum*, the tube is still shorter, and the
flowers are regularly visited by flies and humble-bees.

STELLATÆ.

We have four British genera of this order, Rubia (the Madder),
Galium, Sherardia (Woodruff), and Asperula.

The flowers are small, but in many cases rendered conspicuous by
association. Several of the species are sweet-scented, and attract
insects by means of honey, which is either exposed on a flat dish (Rubia
and Galium), or contained at the base of a short tube (Sherardia and
Asperula). The stamens and pistil ripen simultaneously, and if not
visited by insects, the flowers fertilise themselves. The florets of *Rubia
peregrina* are greenish ; those of *Sherardia arvensis* blue or pink ; the
others either white or yellow. Müller calls attention to the influence of
colour in the case of *Galium mollugo* and *G. verum*, which agree closely
in most points, but the former of which is white, while the latter is
yellow, which he says renders it much more attractive to small beetles.

Fritz Müller has described (*Bot. Zeit.* 1866, p. 129) a very interesting
South American species of this group, *Martha (Prosoqueria) fragans*,
in which the stamens are irritable, and when touched by the proboscis
of an insect, immediately explode, and throw the pollen on to the
insect, at the same time closing the entrance to the tube of the flower,
in which the pistil is situated, and thus preventing the possibility of
self-fertilisation.

VALERIANEÆ.

Of this family we have only one truly British genus, Valeriana,
though *Centranthus ruber*, having been long cultivated in gardens, has
become naturalised in some parts of England.

The flowers of the Allheal (*Valeriana officinalis*), though small, are
rendered conspicuous by association. They are melliferous, and the
honey is accessible even to short-tongued insects, by which they are
much frequented. They are proterandrous.

Valeriana dioica, while agreeing with the preceding as regards the
honey, is, on the contrary, generally diœcious, the male flowers being,
as usual, larger than the female, and, consequently, in most cases
visited first.

COMPOSITÆ.

This great group contains no less than forty British
genera, and a very large number of species. It

comprises the Daisy (Bellis), Dandelion (Taraxacum), Groundsel (Senecio), Chrysanthemum, Thistle (Carduus), Lettuce (Lactuca), Hawkweed (Hieracium), &c. Though there are many differences in the structure of the flowers, as might naturally be expected in so large a group, still in many respects, they are very uniform. The florets are so closely united on a common head, that each group is commonly, though of course incorrectly, spoken of as a single flower. The so-called flower of the Daisy, for instance, is in reality a group of flowers; the outer row of which, or ray florets as they are termed, are unlike the rest and terminate on the outer side in a white leaf or "ray."

The advantages of this arrangement are:—

1. That the flowers become much more conspicuous than would be the case if they were arranged singly.

2. That the facility with which the honey is obtained renders them more attractive to insects.

3. That the visits of the insects are more likely to be effectual, since the chances are that an insect which once alights, touches several, if not many, florets.

No wonder, therefore, that the Compositæ are the most extensive family among flowering plants, are represented in every quarter of the globe and in every description of station (Bentham, "Handbook of the British Flora," vol. i. p. 408; *Jour. Linn. Soc.* 1873, p. 335,) and contain nearly ten thousand species.

The principal differences among the Compositæ, regarded from the point of view of the present work, consist in the different length of the florets, rendering the honey more or less accessible to insects ; in the arrangement of the stamens and pistil ; and in the character of the outer, or "ray" florets.

In some of the Compositæ the florets all contain both anthers and stigma. Generally, however, the ray florets develope no anthers, but a pistil only ; while in some species of *Centaurea* they are barren, and merely serve as flags. It is remarkable that in *C. nigra*, while the outer row of florets generally resemble the rest, they are sometimes enlarged and neuter, as in *C. cyanus*, &c. As regards the relation to insects, we find every gradation, between Taraxacum, *Cirsium arvense*, and Achillea, on the one hand which are conspicuous, rich in honey and much visited by insects ; and on the other, *Senecio vulgaris*, which is rarely visited by insects, and the species of *Artemisia*, which are said to be wind fertilised.

In *Tussilago farfara* the disk florets give honey and pollen ; the ray florets contain neither, but render the flower-head conspicuous, and produce seed.

In the common Feverfew, or large white Daisy (Fig. 86), (*Chrysanthemum parthenium*), which has been well described by Dr. Ogle, "Popular Science Review," April 1870, the flower-heads consist of an outer row of female florets, in which the tubular corolla terminates on the outer side in a white leaf or ray, which doubtless is useful in making the flower conspicuous. The inner florets are also tubular, but are small, yellow, and without a ray. Each of these florets is furnished with stamens as well as a pistil. The anthers are united at their sides so as to form a closed tube, within which the pistil lies. They ripen before the pistil, and open on their inner sides, so that the pollen is discharged into the upper end of the tube above the head of the pistil. When the flower opens, the pollen is already ripe, and fills the upper part of the stamen tube. A floret in this condition is represented in (Fig. 87). The pistil, however, continues to elongate, and at length pushes

Fig. 86.—*Chrysanthemum parthenium.*

the pollen against the upper end of the tube, which gives way, and thus the pollen is forced out of the tube, as shown in (Fig 88). The pistil itself terminates in two branches, which at first are pressed closely to one another, and each of which terminates in a brush of hairs (Fig. 89). As the style elongates this brush of hairs sweeps the pollen cleanly out of the tube, and it is then removed by insects. When the pistil has attained its full length, the two branches open and curve downwards, so as to expose the stigmatic surfaces (Fig. 89 *st*) which had previously been pressed closely to one another, and thus protected from the action of the pollen. From this arrangement it is obvious that any insect alighting on

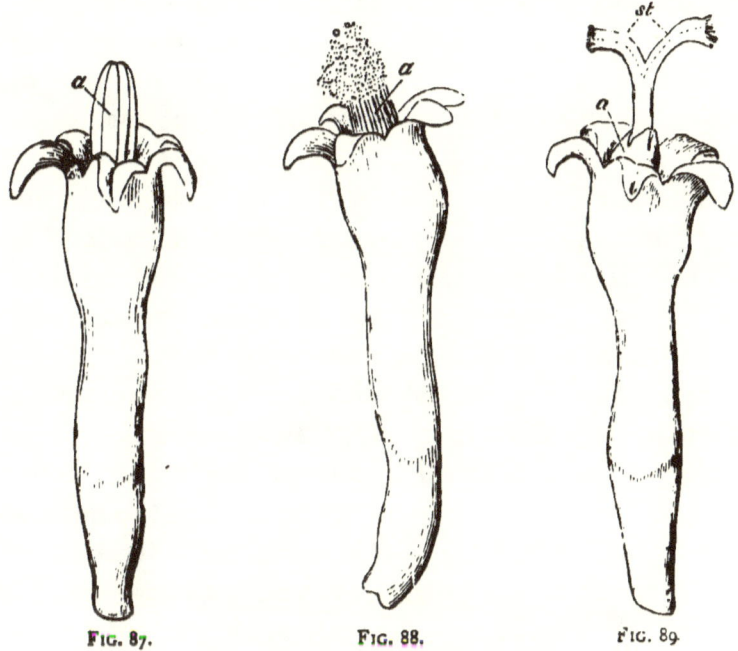

FIG. 87.   FIG. 88.   FIG. 89.

FIG. 87.—Floret of *Chrysanthemum parthenium*, just opened.
FIG. 88.—Ditto, somewhat more advanced.
FIG. 89.—Ditto, with the stigmas expanded.

the flower-head of the Chrysanthemum would dust its under-side with the pollen of the younger flowers, which then could not fail to be brought into contact with the stigmatic surfaces of the older ones. As the expansion of the flowers begins at the outside and thence extends to the centre, it is plain that the pollen of any given floret cannot be used to fertilise one situated on its inner side. Consequently, if the outer row of florets produced pollen, it would, in the great majority of cases, be wasted. I have, however, already mentioned that these florets do not produce pollen, while the saving thus effected enables them to produce a larger corolla. It is also interesting to observe that in these outer flowers the branches of the pistil do not possess the terminal brush of hairs which, in the absence of pollen, would be useless.

In other Compositæ, as in the Marigold, while the ray flowers produce no pollen, the disk flowers develop no stigmas. In this case, as in the Feverfew, the pistil of the ray flowers does not require or possess the terminal brushes of hairs, there being no pollen to be swept out. The central flowers, on the other hand, though they develop no stigmas, require a pistil in order to force the pollen out of the anther tube. Hence the pistil is present as usual, but the head is simple and not bifid. This complete alteration of the function of the pistil is extremely curious.

In *Chrysanthemum leucanthemum* according to Müller, the pistil of the ray florets possesses a terminal brush, which, however, is much less developed than in the disk florets. *Matricaria camomilla* agrees in most respects with Chrysanthemum. The strong smell of this flower, however, seems to be distasteful to bees, though Müller has observed it to be visited by *Prosopis signata* and *Sphecodes gibbus*. It is said to be generally fertilised by flies. Anthemis resembles the two preceding genera in many respects, but differs in possessing scales between all, or at least the central, florets of the receptacle.

The Common Daisy (*Bellis perennis*) has ray florets 1—2 mm. in length, united into a yellow disk 6 mm. in diameter, and surrounded by a row of florets, each terminating in a white "flag" 5 mm. in length. These ray florets are exclusively female, and the pistil has lost the terminal brush of hairs. The two branches are long and clothed on their whole upper surfaces with rows of stigmatic papillæ. The pistil of the ray flowers, on the contrary, has short branches, terminating in a tuft of hairs, and only provided with a small number of stigmatic papillæ. When fertilised, the pistil retires again into the tube of the floret.

In *Inula dysenterica* (the Fleabane) the disk florets contain both stamens and pistil; the ray florets a pistil only, which, however, agrees exactly with that of the disk florets, even in the position of the terminal hairs, which in the absence of pollen, must apparently be useless.

In *Tussilago farfara* the disk florets are male, the ray florets female. In the disk florets the ovary is rudimentary; they contain honey at the base of the tube, which has a length of 4 mm. The pistil terminates in the usual tuft of hairs. The ray florets, on the contrary, produce no pollen; they open, and as the stigmas are mature, before the anther tubes of the disk flowers have opened, they are in fine weather almost always fertilised by the pollen from other flowers.

In the Common Groundsel (*Senecio vulgaris*), 60 to 80 florets are united on one receptacle. The lower, tubular, portion of the floret has a length of $3\frac{1}{2}$ to 4 mm.; the bellshaped portion, only of 1—$1\frac{1}{2}$ mm. The flower heads have no ray flowers, and being therefore much less conspicuous than the allied species, are rarely visited by insects.

Carduus arvensis (Cirsium of some authors) is the commonest of our thistles. Each head contains about 100 florets. The tube of the florets is 8—12 mm. in length, the upper part forming a bell-shaped reservoir 1—$1\frac{1}{2}$ mm. in depth, with five diverging linear lobes. As the lateral florets turn outwards, the whole form a flower head, as much as 20 mm. in diameter. Being therefore very conspicuous, and as the honey in this species and most of its allies rises into the cup of the flower, so as to be accessible even to insects with very short tongues, it is visited by a large number of species. Müller records no less than 88. In *C. lanceolatum*, on the contrary, though it is also a very common species, still in consequence of the cup being somewhat deeper (4—6 mm. against 1—$1\frac{1}{2}$ in *C. arvensis*), and the honey therefore rather less accessible, he only records twelve. In *C. palustris* the depth of the cup is intermediate between those of the two preceding species, and also the number of insect visitors, namely 22.

Onopordon differs from Carduus only in the character of the receptacle, which does not bear chaffy bristles, as in that genus.

The genus Centaurea offers several interesting points. In *C. jacea*, which is sometimes, for instance by Bentham, regarded as a variety of *C. nigra* (the Knapweed), 60—100 florets are united into a head ;

the tubes of the florets are 7—10 mm., the cups 3—4½ in length, each with five long, linear, lobes. The divergence of the outer florets gives the whole head a diameter of 20—30 mm. The hairs constituting the pollen brush are not situated at the extremity of the stigmas as in the preceding species, but form a ring round the pistil at the spot where it bifurcates. When the flower opens the pollen has been already shed into the anther tube in the upper end of which it lies, occupying the space between the anthers and the pistil, and supported by the ring of hairs. If now the flower remains untouched, after a while the stigmatic lobes separate, and some of the pollen falls on them. But if, as generally happens, an insect alights on the flower, or if in any other way the tip of the anthers is touched, immediately the stamens contract, exposing the pollen, which is supported by the stigmatic lobes. Gradually the pistil elongates, and the stigmatic lobes separate ; by which time the pollen has generally been all removed, as the flowers, in consequence of their richness in honey, are much frequented by insects.

In *C. nigra* the outer florets are sometimes of the same size as the rest, sometimes larger, and without either stamens or pistils. In *C. scabiosa* this is always the case. The tubes of the florets also are longer, the cups deeper, and the honey less accessible, in consequence of which it has fewer insect visitors. Müller records only 21 against 48 in *C. nigra.* In *C. Cyanus* also the ray florets are neuter. The contractility of the stamens is very marked. In flowers kept in a room, Müller observed that when touched, they rapidly withdrew themselves 2—3 mm., and then more slowly, 4—6 mm.

Taraxacum (the Dandelion). In *T. officinale* the heads consist of 100—200 florets. In fine weather they stand open, but at night and during rain they close completely. The two lobes of the stigma gradually curl over, so that if the visits of insects are delayed the flower always fertilises itself. The honey, however, is so abundant, and rises therefore so high in the floret, that it is very accessible to insects, no less than 93 species of which have been observed by Müller to visit this plant. The brightness of its colour, the quantity of its honey, the habit of closing in unfavourable weather, and the power of self-fertilization, go far to explain the great abundance of the Dandelion.

The genus *Artemisia* has minute greenish florets, and is said to be wind-fertilised.

This order is the subject of an admirable memoir by Hildebrand (Ueber die Geschlechtsverhältnisse bei den Compositen).

DIPSACEÆ.

There are two British genera belonging to this order; Dipsacus (the Teasel) and Scabiosa. The so-called flower is a compound flower head, as in the Compositæ, from which, however, this group may be at once distinguished by possessing free anthers. Each floret, moreover, is inserted in a small "involucel."

Dipsacus is a proterandrous genus. The pistil terminates in two lobes, the upper surfaces of which constitute the stigma. As, however, in consequence of the stiff spines which radiate on all sides from the flower heads of this plant, the humble bees, by which it is principally fertilised, can only touch the florets with their heads, the two lobes often get in one another's way, and according to Müller it would be a distinct advantage if one of them were absent. He points out also that in fact one of them is sometimes rudimentary, or even occasionally altogether absent. This adaptation then, it would seem, has actually commenced. The leaves form a cup round the stem in which water accumulates, and many small insects are drowned. These it has been supposed contribute to nourish the plant, and Mr. Francis Darwin has observed that protoplasmic filaments extend into the liquid.

Scabiosa arvensis is also proterandrous. About 50 florets are united in one head; they increase in size from the centre towards the circumference, while in *Sc. columbaria* the outer row is considerably larger than the rest, and in *Sc. succisa* they are nearly equal in size. The honey is at the base of the tubular florets, which, however, are more or less funnel-shaped at the mouth, thus greatly facilitating the access of insects. Not only are the florets proterandrous, but this is the case with the whole head; for, though the anthers come to maturity slowly and (as a general rule) successively from the edge to the centre, none of the stigmas emerge until the anthers have all shed their pollen, when they rapidly come to maturity. The male condition of the flower-head lasts several days; the stigmas, on the contrary, come to maturity almost simultaneously. This difference is obviously an advantage. From the length of time during which the anthers are ripening, whenever there is a sunny day, and the insects are abroad, they are almost sure to find some anthers ready to dust them with pollen. On the other hand, the stigmas being mature at the same time, they are capable of being fertilized by a single visit.

Besides the flower-heads with hermaphrodite florets, there are others which contain female florets only, the stamens being more or less rudimentary. This is also an advantage, because if it were otherwise the quantity of pollen would be unnecessarily large. *Scabiosa arvensis* is visited by a great variety of insects belonging to several orders. .

Sc. columbaria has a row of distinctly larger ray florets, while the central ones are all of equal size ; the florets also are smaller than in *Sc. arvensis ;* and consequently, in heads of the same size, more numerous ; the florets appear to be all hermaphrodite ; and the ripening of the anthers does not take place, successively, from the outside.

CAMPANULACEÆ.

The flowers of Campanula are much frequented by insects, and secrete honey at the base of the bell. The anthers are distinct, the filaments of the stamens are expanded at the base into triangular valves which serve to protect the honey ; the pistil is cleft at the top into two, three, or five stigmatic lobes. The genus is widely distributed and contains numerous species.

The accompanying figures show a flower of *C. medium* in three stages. In the bud (Fig. 90) just before opening, it will be seen that the large, long anthers clasp the pistil, which is no longer than they are themselves. In the second stage (Fig. 91) the anthers have opened on the inner side, and shed their pollen, which adheres to the style. The anthers themselves then shrivel up, offering a surprising contrast to their former condition. Insects visiting the flower for

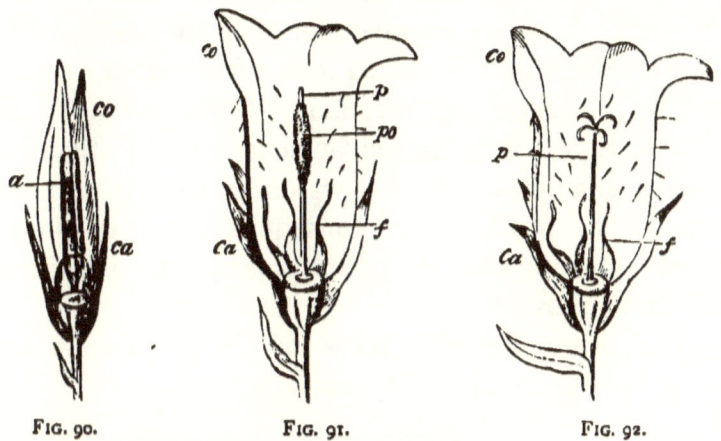

Fig. 90.   Fig. 91.   Fig. 92.

Fig. 90.—Section of bud of *Campanula medium.*
Fig. 91.—Section of a flower in the first (male) condition.
Fig. 92.—Ditto, in the second (female) condition.

the sake of honey, do not, as far as I have observed, generally walk on the petals, being deterred by the stiff hairs which are scattered on their inner surface. In any case, however, they are almost sure, sooner or later, to clasp the style, when they necessarily dust themselves with the pollen. In this stage the flower is incapable of fertilization. Gradually, however, the style elongates, and the lobes of the upper end

K

separate, so that by the time the pollen is all removed
the flower is in the state shown in Fig. 92, and it
is evident that any bee which may have visited a
younger flower, and dusted its under side with pollen,
can hardly fail to deposit some of it on the stigmatic
surfaces thus extended for its reception.

It had been supposed that the hanging position of
Campanula and other bell-shaped flowers had reference
to the position of the stamens and pistil, so that the
pollen might fall from the former on to the latter.
Sprengel, however, pointed out that the real advantage to the flower consisted in the fact that the honey
is thus protected against rain. If the pollen fell on
to the stigma, it is indeed obvious that the stigmatic
surface should be turned upwards, whereas it is at the
end of the pistil, and is consequently turned downwards, showing that the pollen comes from below and
not from above.

The other British genera of Campanulaceæ are
Lobelia, Jasione, and Phyteuma.

ERICACEÆ.

This order contains ten British genera.

Erica tetralix (the Cross-leaved Heath) has been
well described by Dr. Ogle (*Popular Science Review*,
April 1870). The flower is in the form of a bell
(Fig. 93), which hangs with its mouth downwards,
and is almost closed by the pistil, and stigma (*st*),
which represents the clapper. The stamens are eight
in number, and each terminates in two cells, which
diverge slightly, and have at their lower end an oval

opening. But though this opening is at the lower
end of the anther cells, the pollen cannot fall out,
because each cell, just where the opening is situated,
rests against the next anther cell, and the series of
anthers thus form a circle surrounding the pistil and
not far from the centre of the bell. Each anther

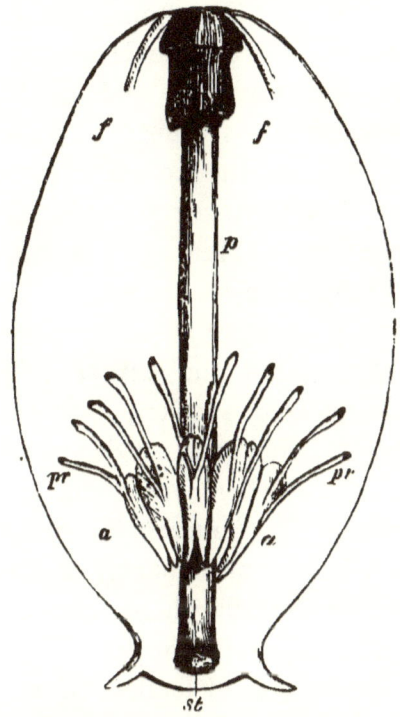

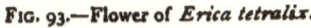

FIG. 93.—Flower of *Erica tetralix.*   FIG 94.—Stamen of ditto.

cell also sends out a long process (*pr, pr*), which thus
form a series of spokes, standing out from the
circle of anthers. Under these circumstances, a
bee endeavouring to suck the honey from the
nectary cannot fail firstly to bring its head in contact
with the viscid stigma (Fig. 93, *st*) and thus to deposit
upon it any pollen derived from a previous visit ;
and secondly, in thrusting its proboscis up the

bell, it inevitably comes in contact with one of the
anther processes, *pr*, which then acts like a lever, and
dislocates the whole chain of anther cells, when a
shower of pollen falls from the open anther cells on
to the head of the bee.

Erica cinerea agrees very closely with *E. tetralix.*
In *Erica* (or *Calluna*) *vulgaris* (the Common Heath),
on the contrary, where the flowers are, in their natural
position, more horizontal, the stamens and pistil
incline upwards, so that insects press their proboscis
under them, and in this manner the pollen is less
likely to be wasted, than if they were central as in *E.
tetralix.* In *Erica vagans* (the Cornish Heath), *E. carnea*,
and *E. ciliata*, the anthers have no appendages.

In the allied genus Vaccinium there is an arrangement similar to that in Erica, but the anther cells
are closed, not by touching one another, but by
resting against the style, so that the style itself
closes the openings until the anthers are disturbed
by the proboscis of the bee. *V. uliginosum* is much
larger than *V. Myrtillus*, and consequently more
conspicuous ; *V. Myrtillus*, on the other hand, has the
compensating advantage of being richer in honey.

The curious, brown-coloured, nearly leafless *Mono-
tropa* (Yellow Birds-nest), differs very much from the
rest of the order.

PRIMULACEÆ.

This order is represented in Britain by eight genera :
Primula, Lysimachia, Trientalis, Glaux, Anagallis,
Centunculus, Samolus, and Hottonia. Cyclamen also
grows wild in some places, but is not a true native.

I have already referred to the genus Primula in the

introductory chapter (*antè*, p. 33). The majority of the species are dimorphous, but not all (Scott, " Proc. Linn Soc.," vol. viii. 1864). In Primula Stricta, according to Axell (" Om Anord. för de Väx Befrucktning "), when the flowers first open, the anthers are already mature, and are attached to the tube of the corolla, some distance above the as yet immature stigma. Gradually, however, the pistil elongates, bringing the stigma to the same height as the anthers.

Hottonia palustris, though so unlike Primula in habit and appearance, is also dimorphous, and agrees · with the former genus very nearly in the relative positions of the stamens and pistil in the two forms. The difference was noticed by Sprengel, who says (p. 103), " I think this is not accidental, but a provision of nature, though I am not in a position to point out the advantage of it."

Lysimachia vulgaris produces no honey. In this species Müller has observed the existence of two extremes (connected, however, by intermediate forms); one, more conspicuous, which rarely or never fertilises itself; the other less conspicuous frequenting shady places, and habitually self-fertile.

Of the genus Anagallis (the Pimpernel) we have, according to Bentham, two species only, *A. arvensis* and *A. tenella*. The former, however, contains two well-marked varieties, one blue and the other red, which do not cross, and are considered by some botanists as distinct species, under the names of *A. cærulea*, and *A. arvensis*. Whether it may be more convenient to treat them as true species or as mere varieties, it must at least be admitted that they differ considerably.

Not only are they of different colou.s, the one blue, the other red, but *A. cærulea* is very decidedly smaller. The stamens and pistil ripen simultaneously. The flowers contain no honey, and partially close about three o'clock in the afternoon.

· The flowers are seldom visited by insects, and it would appear that they generally fertilise themselves. This is said to be the case also with *Centunculus minimus*.

LENTIBULACEÆ.

This order contains two British genera : Utricularia and Pinguicula. Both are fertilised by insects, and in both the insect first touches the stigma, and afterwards comes in contact with the stamens. · In Utricularia the stigma is irritable and retracts at once on being touched, so that the proboscis after dusting itself with the pollen does not again come into contact with it.

Both genera are insectivorous. Utricularia is aquatic, and the submerged leaves bear small bladders or utricles, at the entrance of which are stiff hairs so arranged as to permit the entrance, but prevent the exit of small water animals. Even fish, of course only when very young, are sometimes so captured. ·

In Pinguicula the leaves are covered with sticky, glandular hairs, and the escape of any small flies or other insects which may be so unfortunate as to alight on them is rendered more difficult by the fact that the edges are curved over.

APOCYNACEÆ.

In Vinca (the Periwinkle), which has been described by Delpino and Hildebrand, the arrangement resembles in principle that already described in

Polygala. The anthers and the stigma, which is immediately below them, together nearly close up the tube of the flower. The upper portion of the pistil is clothed with hairs which arrange themselves so as to form a sort of pocket or chamber opposite each anther, and when the pollen is shed it is received into this pocket or chamber. The stigma somewhat resembles an inverted saucer, attached by the middle to the style. The upper portion of the stigma is viscid and rubs against the proboscis of the insect as it is withdrawn. The proboscis, thus rendered adhesive, carries off some of the pollen. When the insect visits the next flower, the pollen is scraped off the proboscis by the sharp edge of the saucer, and is thus accumulated in the hollow of the saucer, which is the true stigmatic surface.

GENTIANACEÆ.

In this order we have six British genera : Cicendia, Erythræa, Gentiana, Chlora, Menyanthes, and Limnanthemum.

Gentiana Pneumonanthe is proterandrous. It secretes honey at the base of a tube 25—30 mm. long ; Bees, however, can creep half way down, in doing which they come in contact with the anthers in younger flowers, and in older ones with the stigma, which lies somewhat higher in the tube. The power of self-fertilisation appears to be lost. *Gentiana amarella*, on the contrary, is homogamous, the anthers and stigma coming to maturity together, though as the style of pistil is somewhat longer than the stamens, an insect touches the stigma before reaching the anthers.

The beautiful *Erythræa centaurium* is frequently

visited by butterflies, though it contains no honey, at least neither Sprengel nor Müller could find any. Menyanthes and Limnanthemum (Kuhn, " Bot. Zeit.," 1867) are said to be dimorphous.

POLEMONIACEÆ.

. This family is represented in England by one species, *Polemonium cæruleum,* and even this is a doubtful native. It has been shown by Axell to be proterandrous.

BORAGINACEÆ.

This order is easily distinguished from all others, except the Labiatæ, by the four seed-like nuts ; from the Labiatæ by the form of the flowers, and by the leaves being alternate. It contains eleven British genera, viz.,—Echium, Pulmonaria (Fig. 96), Mertensia, Lithospermum, Myosotis, Anchusa, Lycopsis, Symphytum, Borago (Fig. 95), Asperugo, and · Cynoglossum.

In consequence of its conspicuousness, and the easy accessibility of its honey, *Echium vulgare* is visited by a great variety of insects. The flower is tubular, contracting towards the base, so that insects are naturally conducted to the honey. The stamens are five in number ; one remains in the tube of the flower, while the other four project, and form a con- venient alighting stage for insects, which can thus hardly fail to dust their undersides with pollen.

Echium is proterandrous ; when the flower opens the anthers are already ripe ; the pistil, on the other hand, is still quite short and immature, scarcely reaching to the mouth of the tube. Gradually, how- ever, it extends until it reaches 10 mm. beyond

the tube, and divides at the end into two short
branches, with terminal stigmas. In this species,
therefore, cross-fertilisation is favoured; firstly, by
the fact that the stamens ripen before the stigmas;
and, secondly, by the relative position of the two, the
stigmas, as we have seen in so many other cases,
projecting somewhat beyond the stamens. Under
these circumstances cross-fertilisation is so thoroughly

FIG. 95.—*Borago officinalis.*

secured, that the plant is said to have lost the
power of fertilising itself. Müller observed no less
than 67 species of insects on the flowers of this
plant: some of which (*Osmia adunca* and *O. cœmen-
taria*) seem to confine themselves to it.

In the Borage (*Borago officinalis*, Fig. 95) we find
an arrangement of the stamens and pistils very

different to that in Echium, but, as Sprengel has pointed out, somewhat resembling that already described in the Violet. The flowers are drooping, of a beautiful blue, with a white central circle ; dark stamens, combined into a tube, and a pink pistil. The pale yellow, fleshy ovary secretes honey, which lies in a short tube formed of the basis of the stamens. The anthers are long, and open gradually from the apex to the base, so that the pollen falls into the closed space between them and the pistil. This arrangement effectually protects both the pollen and the honey from all insects, excepting bees. The latter, however, force their proboscis down to the honey, between the anthers, which, however, return to their former position again, as soon as the proboscis is withdrawn. As soon as the anthers are separated, the pollen drops down on to the head of the bee, and is thus carried from one flower to another. Cross-fertilisation is also favoured by the flower being proterandrous, the stigma not becoming mature until the anthers have shed all their pollen. The Borage is much visited by bees, especially by the common hive bee.

Pulmonaria officinalis (Fig. 96) is a dimorphous species ; being rich in honey and much visited by insects, it has not only lost the power of self-fertilisation, but is said by Hildebrand (*Bot. Zeit.*, 1865) to be sterile to pollen from the same form of flower; that is to say, long-styled flowers require to be fertilised by pollen from short-styled flowers, and *vice versâ*. Darwin, however, succeeded in obtaining seeds and raising seedlings from some long-styled plants which were fertilised with pollen of the same form. (*Jour. Linn. Soc.*, v. x. p. 430.) We have already seen that this is partially the case with other dimorphous species.

The genus Myosotis (the Forget-me-not) has already been alluded to in the introductory chapter (*ante*, p. 35). The species, however, appear to differ among themselves in the relative positions of the stamens and pistil.

FIG. 96.—*Pulmonaria officinalis.*

In this beautiful and interesting family, though we have not above twenty British species, we find, as Müller has well pointed out, the widest differences in the conditions of fertilisation. *Pulmonaria officinalis* is dimorphous, and sterile—not only with its own pollen, but even in some cases with that of a different flower, unless it belongs to the different form. *Echium vulgare* has lost the power of self-fertilisation, but, so

far at least as we know, is fertile with the pollen of
any other flower belonging to the species.   Other
species are generally fertilised by insects, but in their
absence perform this office for themselves; while
lastly, some species, such as *Lithospermum arvense*,
and *Myosotis intermedia*, habitually fertilise themselves.   Again cross-fertilisation is secured in Pulmonaria by dimorphism; in Echium and Borago by
proterandrousness (if I may be permitted to coin the
word) : in Symphytum and Anchusa, by the projection of the stigma beyond the stamens; in Lithospermum and Myosotis, by the narrowness of the
flower tube.

CONVOLVULACEÆ.

The well-known Convolvulus and the singular little
Dodder (Cuscuta) are the only British genera belonging to this family.

Cuscuta is a leafless, annual, parasitic plant, with
thread-like stems.   The flowers are small, nearly
globular, and grow in lateral heads or clusters.   One
species attacks the clover, and is sometimes sufficiently
abundant to do much mischief.

We have in England three species of Convolvulus
—*C. arvensis*, *C. sepium*, and *C. soldanella*.

C. arvensis being melliferous and slightly sweet-
scented, is much visited by insects.   The honey is
situated below the bases of the stamens, which are
somewhat flattened and bent inwards, so that the
insect can only reach the honey by pressing its proboscis down between them.   The stigmas and anthers

mature at the same time ; but as the former project above the latter, they are necessarily touched first. If the visits of insects be too long deferred, the flower fertilises itself. *C. arvensis* closes in wet weather and at night.

C. sepium, on the contrary, remains open during rain, but closes at night, unless there be a moon, when it remains expanded. It has no smell, and is perhaps, on that account, in spite of its large size, comparatively little visited by insects.

SOLANACEÆ.

The British genera are the following : Hyoscyamus (the Henbane), Solanum (the Nightshade), and Atropa. Datura is sometimes found growing wild, but it is not a true native.

Solanum secretes no honey, and is little visited by insects. Hyoscyamus, on the contrary, is melliferous, and cross-fertilisation is favoured by the projection of the stigma beyond the anthers.

OROBANCHACEÆ.

A curious family, with simple or rarely-branched stems, and scales instead of leaves. The species are either brown or purplish, but never green, and are parasitical on the roots of other plants. There are two British genera : Orobanche (Broomrape) and Lathræa ; both are parasitic. In Lathræa the scale-like leaves are hollowed out, the inner surfaces being provided with peculiar structures of two kinds ; both consist of three cells, two of which are spherical, and situated on the third, which in the one sort is cylindrical, so that they resemble glandular hairs ; in the second sort the basil cell is flattened. These organs have been described as possessing the power of throwing out protoplasmic extensions, but this has not yet been confirmed.

SCROPHULARIACEÆ.

This is a large family, consisting of fourteen genera, and contains : Veronica (Fig. 97), Verbascum (Mullein), (Fig. 98), Linaria, Antirrhinum (Snapdragon), Scrophularia (Fig. 99), Digitalis (Foxglove), (Fig. 100), Euphrasia (Eyebright), (Fig. 106), Rhinanthus (Rattle), &c.

The first two genera have more or less open flowers; while the others are more distinctly tubular, and have much the appearance of Labiatæ, but differ from that group in having the ovary two-celled, with several ovules in each cell.

Veronica. The flowers are rendered conspicuous by their colour and the association in racemes. In *V. Chamædrys* (Fig. 97), the anthers and stigmas ripen simultaneously, but while the latter project

FIG. 97.—*Veronica Chamædrys.*

straight forwards, the two stamens turn outwards, so that fertilisation can hardly take place.

V. Beccabunga in many respects resembles *V. Chamædrys*; but is proterogynous. In *V. spicata* some flowers are proterogynous, others proterandrous, and being, in consequence of their conspicuousness, much visited by insects, they appear to have lost the power of self-fertilisation. In *V. hederæfolia*, on the contrary, the flowers are minute, and habitually fertilise themselves.

The species of Verbascum (Mullein) are showy plants, with either white or yellow flowers, forming a tall spike, which in *V. Thapsus* reaches a height of four feet. *V. nigrum*, L. (Fig. 98) has yellow flowers ; the stamens clothed with beautiful violet hairs. They secrete very little honey, but are visited by various

FIG. 98.—*Verbascum Thapsus.*

insects for the sake of the pollen, and perhaps also on account of the glandular terminations of the violet staminal hairs. The stamens turn somewhat upwards, the pistil, on the contrary, downwards, so that an insect alighting on the lower lip of the corolla, which is the most convenient place, would naturally come in contact with it before touching the stamens. *V.*

nigrum, however, according to Gærtner, cannot be fertilised by its own pollen.

The genus Scrophularia, from which the family takes its name, is remarkable in many respects. From the general arrangement of the blossom in flowers of the Labiate form, the pistil could hardly occupy any other position than the central median

FIG. 99.—*Scrophularia nodosa.*

line, and a fifth stamen would accordingly be in the way. It has therefore disappeared, though Müller mentions that he once found one in *Lamium album*. In *Scr. nodosa* (Fig. 99), however, the four normal stamens and the pistil occupy the lower side of the flower, and the presence of a fifth stamen, even if useless, is under these circumstances not injurious. A rudimentary fifth stamen is, in fact, habitually present, and in some cases bears pollen. *Scr. nodosa* is proterogynous, and is much frequented and fertilised by wasps. Pentstemon also has a fifth stamen, which curves in a very curious manner from the upper to the under side of the flower so as to be out of the way of the pistil. Ogle regards it as perfectly useless (*Popular Science Review*, Jan. 1870), but it is so large that I cannot help thinking it must

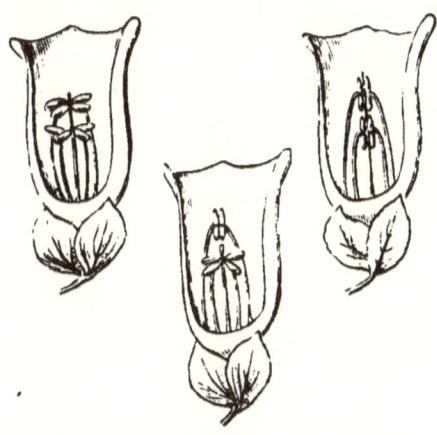

FIG. 100.   FIG. 101.   FIG. 102.

FIG. 100.—Section of *Digitalis purpurea*, showing the anthers unripe and horizontal.
FIG. 101.—Ditto, more advanced. The upper anthers ripe and vertical, the lower ones as before.
FIG. 102.—Ditto, still more advanced. All the anthers ripe and vertical.

have some function, though I am unable to suggest one.

In *Linaria vulgaris* the flowers form a closed box terminating behind in a spur, 10—13 mm. in length, which contains the honey, and the orifice of which is protected by hairs. Under these circumstances, the long-lipped bees are the only insects which can suck the honey. *Antirrhinum majus* (the Snapdragon) differs in the larger size of the flowers, the greater

L

firmness with which they are closed, and in the position of the honey, which lies at the basis of the corolla, and does not penetrate into the short spur, which is hairy, and therefore not suited for such a purpose. They are almost always fertilised by humble bees, though smaller bees occasionally force their way into them.

Digitalis purpurea (the Foxglove) is also exclusively fertilised by humble bees, which alone are large

FIG. 103.—*Bartsia odontites.*

enough to fill the bell, and thus to deposit pollen on the stigma. The flower is proterandrous, but appears to be self-fertile if the visits of humble bees are delayed or prevented. The cells of the anthers, as Ogle has pointed out, are at first transverse (Fig. 100), but as the two pairs ripen they successively become longitudinal (Figs. 101 and 102).

The other British genera of this group have narrow tubular flowers; in which the upper lip protects the anthers and pistil, while the lower lip serves as an alighting stage for insects. The stamens are so arranged that the insects in searching for the honey dust themselves with the pollen. For instance, in *Bartsia odontites* (Fig. 103), the common red Bartsia, the flower forms a tube 4—5 mm. long; at the base of which is the honey, while the entrance is protected against rain by the four hairy anthers. These lie

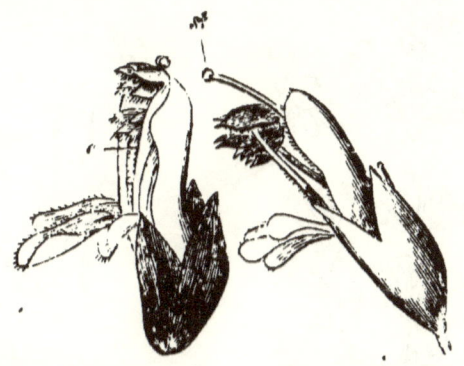

FIG. 104.—*Bartsia odontites.* Flower with a short pistil.   FIG. 105. Ditto. Flower with a long pistil.

close together; but immediately below them, the filaments of the stamens separate so as to leave a space through which bees can insert their proboscis, and thus reach the honey. In doing so they naturally dust themselves with pollen, some of which they transfer to the stigma (Fig. 105, *st*) of the next flower they may visit. Müller has observed that in plants of this species which live in shady places and are consequently less visited by insects, the pistil is shorter (Fig. 104), the stigma consequently nearer to the anthers, and more likely to be fertilised directly by them.

He also observes that this flower is not perfectly adapted to present circumstances, since bees are able to, and often do, insert their proboscis above the stamens, in which case they do not fertilise the flower.

Euphrasia officinalis (the Eyebright) (Fig. 106), agrees in many respects with the preceding; but there

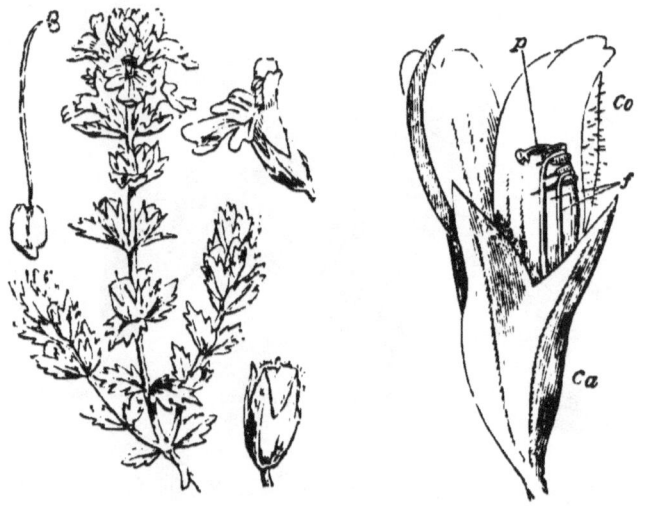

FIG. 106.—*Euphrasia officinalis.*   FIG. 107.—Flower of *Euphrasia officinalis.*

is no room above the stamens for the proboscis of the bee. The anthers (Fig. 107) also, which in *Bartsia odontites* are merely locked together by hair, in this species are more intimately connected, the two uppermost anthers to one another, the lower anther of each upper pair with the upper anther of the lower stamen on the same side. The lower anther of the lower stamen is produced into a strong point (Fig.

107, which is touched by the proboscis of the bee
in passing down the tube to the nectary, and serves
as a lever, shaking the whole system of anthers and
thus causing the pollen to fall out on to the bee.

In this species also H. Müller has observed that
there are two forms, a larger one which is adapted to
be fertilised by insects, and a smaller one which more
frequently fertilises itself.

In *Rhinanthus Cristagalli* (the Common Rattle) the
anthers are locked together, and the pollen is shed
on to the bee, but the mode in which this is effected
is not the same. In this species, as in *Bartsia odontites*, the bee has to pass its proboscis between the
filaments of the anthers in order to reach the honey,
and the space between them is so narrow, that the
bee in pressing its proboscis down the tube, presses
the filaments apart, thus shaking the anthers, and
freeing the pollen. In this species also H. Müller
has observed the existence of two forms.

In the common Pedicularis (Fig. 108) (*Pedicularis
sylvatica*), which has been well described by Hildebrand and Delpino, the arrangement is somewhat different. The anthers open on their inner sides, and
the edges of the open anther cells on the one side of
the flower exactly correspond with, and are applied
to, the corresponding edges of the anthers on the
other side of the flower; each pair of anthers thus
forming as it were, a closed box. The outer sides
of the anthers are slightly attached to the walls of
the hood. But the sides of the hood are somewhat
too near together to admit the head of the humble-
bee, and the insect therefore, in order to reach the

honey, pushes them a little apart, thus opening the
anther-box and letting down a little shower of pollen,
which is prevented from spreading by the fringe of

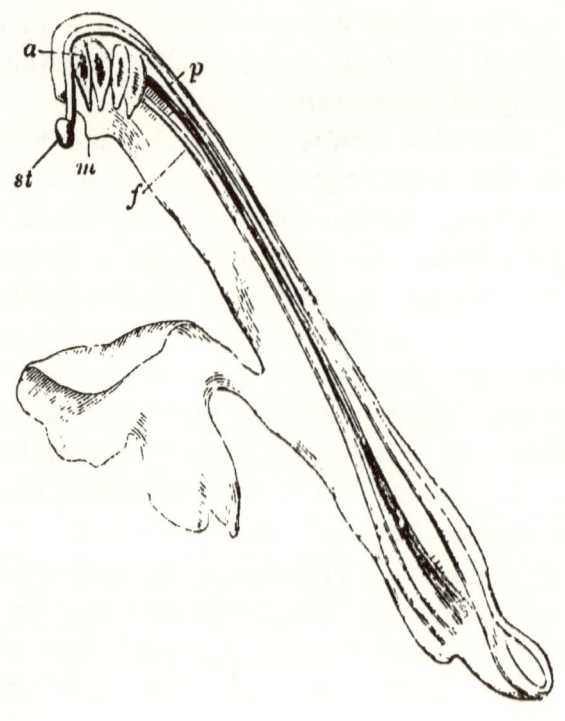

FIG. 108.—*Pedicularis sylvatica.*

hairs on the lower edge of the anther, and thus falls
on to the head of the bee, at the very spot which a
moment before had touched the stigma, and which
will again touch that of the next flower she visits.

In *P. palustris* the point *m* is elongated, and the anthers, in the specimens which I have examined, are glabrous.

The structure of Melampyrum agrees in essentials with that of Pedicularis. In *Calceolaria pinnata*, Hildebrand describes an arrangement somewhat similar to that which we shall meet with in Salvia.

LABIATÆ.

This large and interesting order contains eighteen British genera, amongst which are the Salvia, Dead Nettle, Sage, Thyme, Mint, Marjoram, Bugle, and Calamint. Most of them, if not all, produce honey at the base of the ovary.

In few flowers is the adaptation of the various parts to the visits of insects more clearly and beautifully shown than in the common white Dead Nettle (*Lamium album*), (Fig. 109).

The honey occupies the lower contracted portion of the tube, and is protected from the rain by the arched upper lip and by a rim of hairs. Above the narrower lower portion the tube expands, and throws out a broad lip (Fig. 111 *m*), which serves as an alighting place for large bees, while the length of the narrow tube prevents the smaller species from obtaining access to the honey, which would be injurious to the flower, as it would remove the source of attraction for the bees, without effecting the object in view. At the base of the tube, moreover, at the point marked *ca*, Fig. 111, there is a ring of hairs which prevent small insects from creeping down the tube and so getting at the honey. Lamium, in fact, like so many of our other wild flowers, is especially adapted for humble bees. They alight on the lower lip, which projects at the side, so as to afford them a leverage, by means of which they may press the proboscis down the tube to the honey ; while, on the other hand, the arched upper lip, in its size, form, and

Fig. 109.—*Lamium album.*

position, is admirably adapted not only as a protection against rain, but also to prevent the anthens (Fig. *a, a*) and pistil (Fig. 111, *st*) from yielding too easily to the pressure of the insect, and thus to ensure that it should press the pollen which it has brought from other flowers against the pistil.

The stamens do not form a ring round the pistil, as is so usual. On the contrary, one stamen is absent or rudimentary, while the other four lie along the outer arch of the flower, on each side of the pistil. They are not of equal length, but one pair is shorter than the other; the inner pair in some species, the outer pair in others being the longest.   Now, why is this?   Probably, as Dr. Ogle has suggested, because if the anthers had lain side by side, the pollen would have adhered to parts of the bee's head which do not come in contact with the stigma, and would therefore have been wasted; perhaps also partly, as

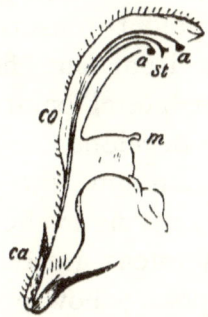

Fig. 110.—Flower of *Lamium album*.   Fig. 111.—Section of ditto.

he suggests, because it would have been deposited on the eyes of the bees, and might have so greatly inconvenienced them as to deter them from visiting the flower.   Dr. Ogle's opinion is strengthened by the fact that there are some species, as for instance the Foxglove, in which, as shown in Figs. 100—102, the anthers are transverse when immature, but become longitudinal as they ripen.

But to return to the Dead Nettle.   From the position of the stigma which hangs down below the anthers (Fig. 111 *st*), the bee comes in contact with

the former before touching the latter, and consequently generally deposits upon the stigma pollen
from another flower. The small processes (Figs.
110, 111 *m*) on each side of the lower lip are the
rudiments of the lateral leaves with which the ancestors of the Lamium are provided. Thus, then,
we see how every part of this flower is either—like
the size and shape of the arched upper lip, the relative position of the pistil and anthers, the length
and narrowness of the tube, the size and position
of the lower lip, the ring of hairs, and the honey
—adapted to ensure the transference, by bees, of
pollen from one flower to another; or, like the
minute lateral points (*m*), an inheritance from more
highly-developed organs of ancestors. If we com-
pare Lamium with other flowers we shall see how
great a saving is effected by this beautiful adaptation.
The stamens are reduced to four, the stigma almost
to a point; how great a contrast to the pines and
their clouds of pollen, or even to such a flower as
the Nymphæa, where the visits of insects are secured, but the transference of the pollen to the
stigma is, so to say, accidental. Yet the fertilisation of Lamium is not less effectually secured than
in either of these.

Lamium maculatum has a somewhat longer tube
(15—17 mm.) than *L. album*, and only bees with a
long proboscis can therefore suck it. *B. terrestris*,
however, obtains access to it by force, and *B. rayellus*,
according to H. Müller, uses the holes made by *B.
terrestris*. In *L. purpureum* the tube is somewhat
shorter.

Lamium amplexicaule, in addition to the normal flowers, also produces cleistogamous ones (Figs. 36, 37), which appear in the early spring and again in autumn.

In this genus it would appear, as already mentioned, that the pistil matures as early as the stamens, and that cross-fertilisation is obtained by the relative position of the stigma, which, as will be seen in the figure, hangs down slightly below the stamens, so that a bee bearing pollen on its back from a previous visit to another flower would touch the pistil and transfer to it some of this pollen, before coming in contact with the stamens.

In other species belonging to the same great group (*Labiatæ*) cross-fertilisation is secured by the fact that the stamens come to maturity, shed their pollen, and are shrivelled up, before the stigma is mature. The genus Salvia was described by Sprengel, and more recently by Hildebrand and Ogle (*Pop. Sci. Rev.* July, 1869). Fig. 112 represents a young flower of *Salvia officinalis* in which the stamens (f) are mature, but not the stigma (p), which, moreover, from its position is untouched by bees visiting the flower, as shown in Fig. 113. The anthers, as they shed their pollen, gradually shrivel up; while on the other hand the pistil increases in length and curves downwards, until it assumes the position shown in Fig. 114 st, where, as is evident, it must come in contact with any bee visiting the flower, and would touch just that part of the back on which pollen would be deposited by a younger flower. In this manner self-fertilisation is effectually provided against. There are, however, several other points in which *S. officinalis* differs greatly from the species last described.

The general form of the flower indeed is very similar. We find again that, as generally in the

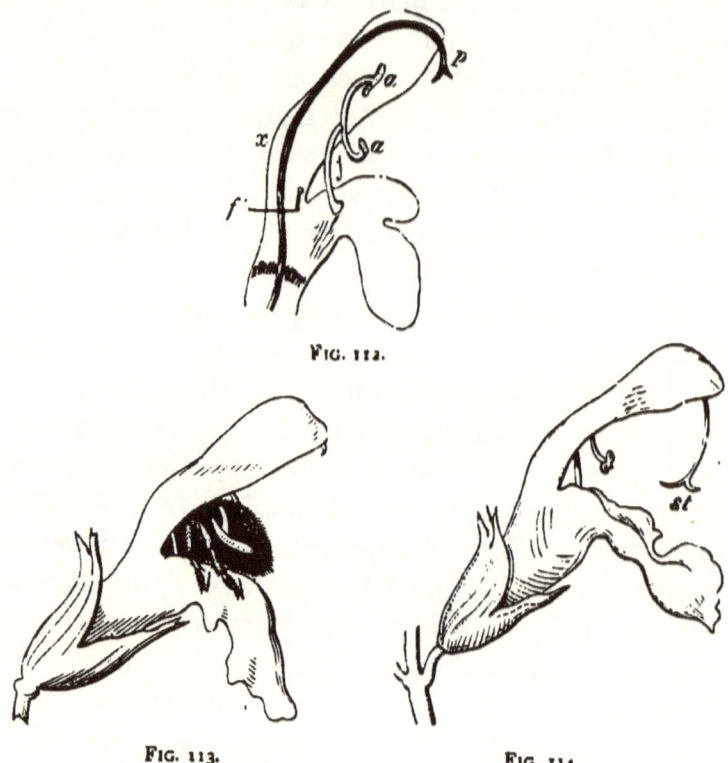

FIG. 112.

FIG. 113.   FIG. 114.

FIG. 112.—*Salvia officinalis.* Section of a young flower.
FIG. 113.—Ditto, visited by a Bee.
FIG. 114.—Ditto, older flower.

Labiates, the corolla has the lower lip adapted as an alighting board for insects, while the arched upper lip covers and protects the stamens and pistils.

In *Salvia officinalis*, however, the back of the

upper lip shows an arch at the part *x*, and the
front portion of the lip, containing the stamens, is
loftier than in Lamium, and does not therefore come
in contact with the back of the bee (Fig. 112). In evident correlation with this arrangement, we find a very
remarkable difference in the stamens (Figs. 115–16).
Two of the stamens (Fig. 112, *f'*) are minute and
rudimentary. In the other pair, the two anther cells
(Fig. 115 *a, a'*) instead of being, as usual, close

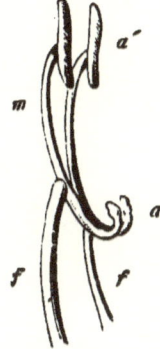

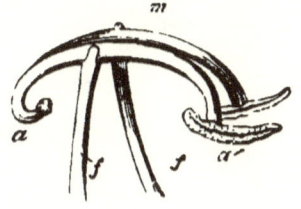

FIG. 115.—Stamens in their natural
position.

FIG. 116.—Stamens when moved by
a Bee.

together, are separated by a long connective (*m*).
Moreover, the lower anther cells (*a, a*) contain very
little pollen; sometimes indeed none at all. This
portion of the stamen, as shown in Fig. 112, hangs
down and partially stops up the mouth of the
corolla tube. When, however, a bee thrusts its head
into the tube in search of the honey, this part of
the stamen is pushed into the arch (*x*), the con-
nectives of the two large stamens revolve on their
axis, and consequently the fertile anther cells (*a'*)

are brought down on to the back of the bee as shown in Fig. 113.

In *S. pratense* the lower branch of the anther is comparatively short. The different species of Salvia differ indeed considerably from one another in this respect. One of them, *S. cleistogama*, produces cleistogamous flowers, as its name denotes.

Teucrium Scorodonia is very markedly proterandrous. When the flower first opens the stigma stands behind the stamens (Fig. 117) and is not touched by

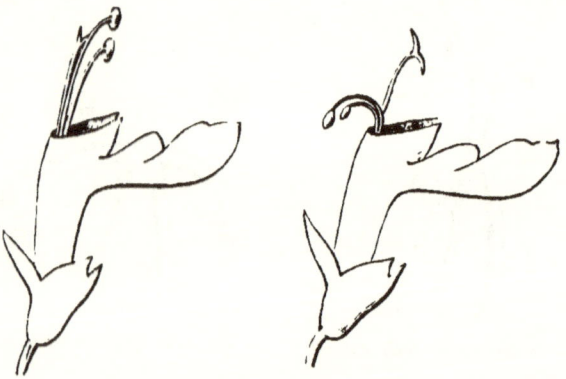

FIG. 117.—*Teucrium Scorodonia*, in the first state.   FIG. 118.—Ditto, in the second state.

the insect. Gradually, however, the stamens turn backwards, and the pistil moves forwards (Fig. 118), so that in older flowers, it stands where the stamens stood before, and in its turn comes in contact with the insect. This flower, though not conspicuous, is a favourite with insects.

In *Ajuga reptans* the upper lip is very short, but the flowers stand close to one another, and the stamens and pistil of each are protected by the lower bract of the flower above. According to Delpino, Ajuga is proterandrous. The pistil is already mature when the flower opens, but then lies behind and is protected by the stamens. After a while the stamens separate a little, so that the stigma is in its turn exposed. In *Ballota nigra* the arrangement of the stamens and pistils is somewhat similar, and the flower is also slightly proterandrous.

In *Galeobdolon luteum*, the flower tube is eight mm. (but, as the upper end is dilated, practically only six mm.) in length. Though the stigmatic ends of the pistil diverge shortly after the opening of the flower, and appear to be then already mature, still they occupy a more prominent position at a later period. In this respect, therefore, it is intermediate between Lamium and Ballota.

Galeopsis tetrahit is a variable plant, and the tube varies in length in different specimens from 11 to 17 mm.; of which, however, the 4—6 upper millimetres are somewhat expanded. This variability is an interesting fact in relation to the theory of natural selection. The pistil, when mature, moves forward, as in the preceding species. *G. ochroleuca* agrees very closely with *G. Tetrahit*, but the tip of the pistil, instead of lying between the anthers of the two longer stamens, projects slightly beyond them. *G. versicolor* has a longer tube, while *G. Ladanum* has a somewhat shorter one; in most respects, however, they agree with *G. ochroleuca.*

Stachys sylvatica is distinctly proterandrous, but has not lost the power of self-fertilisation. In *S.*

palustris the tube is shorter than in *S. sylvatica ;* the
four stamens are of equal length; and when the flower
opens, the anthers of the outer ones lie in front of
the inner ones. When they have shed their pollen
they turn outwards, thus exposing the inner ones,
which in their turn shed their pollen, and then
move outwards to make room for the pistil, which
thus occupies the place which they had previously
filled.

Betonica officinalis is also proterandrous; the pistil
being comparatively short when the flower first opens,
and not attaining its full length until the anthers
have shed their pollen.

In *Calamintha Clinopodium* the upper process of the
stigma varies considerably in size. The stamens are
still more remarkable in this respect, presenting variations which, as mentioned in the case of *Galeopsis
tetrahit,* are very interesting.

I have already in the introductory chapter referred
to the Thyme (*Thymus Serpyllum,* Figs. 32 and 33)
as a type of a proterandrous flower. It is extremely
rich in honey, much frequented by insects, and,
according to Müller, has lost the power of self-
fertilisation. Besides the ordinary flowers, which
contain both stamens and pistils, there are other
smaller ones, which contain a pistil only. In Italy,
Delpino has observed not only these two kinds, but
also a third in which the pistil is quite rudimentary.
Ogle also in England has observed that in some
flowers the pistil never becomes fully developed. On
the contrary in Germany, Hildebrand, Ascherson, and
Müller, have sought in vain for these male flowers.

This geographical differentiation, if it really exist, is very interesting.

H. Müller attempts to explain the presence of these small flowers by pointing out that where there is any variation in the size of the flowers, the smaller and less showy ones would be the last to be visited by the insects. Under these circumstances, as such flowers would be fertilised by the pollen derived from previous visits, the stamens of such smaller flowers would be useless, and would tend to become rudimentary. Further observations are, however, I think, required before this explanation can be regarded as satisfactory.

The Mint (*Mentha arvensis*) is also proterandrous, and, like the Thyme, possesses, in addition to the hermaphrodite flowers, others which are smaller and merely female. Some species of the genus are dimorphous. The genus Mentha seems to be in some respects a connecting link between the typical Labiates, and the ordinary tubular form.

Origanum vulgare (the Marjoram) also has plants with large, proterandrous, bisexual flowers ; and others with smaller female ones. In the secretion and position of the honey it agrees with the Thyme ; but while on the one hand it is less sweet, it is, on the other, more conspicuous. These two differences nearly counterbalance one another ; the flowers are consequently much visited by insects, and have also lost the power of self-fertilisation.

Nepeta glechoma (the Ground Ivy), like the three preceding genera, is proterandrous, and has small

M

female flowers, as well as the larger hermaphrodite ones.

Prunella vulgaris also has the two kinds of individuals, but the female plants are comparatively rare. Axell says that, in the absence of insects, the larger flowers fertilise themselves, but this was not the case with those observed by Müller. If Prunella be really self-fertile this would constitute an argument against Müller's view of the origin of the small female flowers.

Lycopus Europæus is distinctly proterandrous. In this species, as in Salvia, two of the stamens are rudimentary. This is an advantage in Salvia, on account of the curious mechanical structure of the stamens. In Lycopus, the diminution is perhaps connected with the smallness of the size of the flower. Veronica, which has the smallest flowers of all the Scrophulariaceæ, has also only two stamens instead of four, or more.

VERBENACEÆ.

The common *Verbena officinalis* is the only British species of this order. The calyx is five-toothed, the corolla distinctly tubular, and with five somewhat unequal lobes. The stamens are sometimes two, sometimes four, in number. It secretes honey at the base of the tube.

PLUMBAGINEÆ.

There are two British genera of this order, viz. Statice and Armeria. The genus Plumbago has already been referred to in the introductory chapter (*antè*, p. 10) as an illustration of an insect-fertilised flower, in contrast with *Plantago major*, which is wind-fertilised.

PLANTAGINEÆ.

This order contains two British genera ; Plantago and Littorella.

Plantago, the common Plantain, has small, hermaphrodite flowers in heads or spikes on a leafless peduncle. The sepals are four; the corolla has four lobes; the stamens are four, alternating with the petals, and very long; the style is long, simple, and hairy. This genus offers several interesting peculiarities.

Plantago major is proterogynous, and according to Axell, as I have already mentioned (*antè*, p. 10), is wind-fertilised, which, however, is not invariably the case in other species.

In *Pl. lanceolata*, Delpino has observed three different forms :—

Firstly, a form with a strong and high stalk ; white and broad anthers. This he says is entirely wind-fertilised.

Secondly, one with a less elevated stalk, and less exclusively anemophilous. On it he observed a species of Halictus, which endeavoured to collect pollen. The plant is, however, so little suited to this, that most of the pollen fell to the ground.

Thirdly, a dwarf variety, with shorter stamens. This form was visited by several species of bees and is intermediate between wind-fertilisation and insect-fertilisation. Müller also has observed two varieties of this species ; one tall and long-eared, the other shorter and smaller ; both of them were visited by insects. *P. lanceolata* is proterogynous.

M 2

Plantago media is also proterogynous, though less so than *P. lanceolata.* It is more frequently visited by insects, having a slight scent, and stamens with pink filaments. Nevertheless, it appears to be generally fertilised by wind.

According to Darwin, several North American species are dimorphous (*Proc. Linn. Soc.* v. vi., 1862, p. 95), and Kuhn states that some have also cleistogamous flowers.

FIG. 119.—CHENOPODIUM BONUS-HENRICUS.

CHAPTER VI.

INCOMPLETÆ.

Of this sub-class we have in Britain representatives of fifteen orders, some of them very numerous and important. To it, for instance, belong many of our forest trees, such as the elm, oak, beech, birch, poplar, willow, pine, fir, &c. ; and a large number of the common herbs, such as the nettles, chenopodiums, euphorbias (spurges), &c. The flowers, however, are generally less conspicuous (see Fig. 119) than those we have hitherto been considering, and offer fewer adaptations in relation to insects ; being in many cases wind-fertilised: thus in H. Müller's work, less than ten pages are occupied by this whole sub-class, of which more than half are devoted to the Polygonaceæ, and a greater part of the remainder to the Aristolochiaceæ; two orders which in many respects form a marked contrast to the remainder, and have, at least in some species, conspicuous flowers. In the other orders, on the contrary, the flowers are generally minute. Thus in the Paronychiaceæ, Bentham says, "Petals usually none, or represented by five small filaments;" in Santalaceæ, "flowers small;" in Empetraceæ, flowers "minute, axillary;" in Callitrichineæ, flowers "minute;" in Urticaceæ, flowers "small;" in Ulmaceæ, flowers "small;" while in the Amentaceæ (beech, oak, birch, &c.), and Coniferæ, the flowers rarely are coloured, or contain honey. Indeed, it is, I think, a strong argument in favour of Sprengel's views, that while large flowers are almost always coloured, small ones are usually greenish; thus out of thirty-nine British genera of Incompletæ, by far the greater number of which have small flowers, in no less than thirty-seven genera they are also more or less greenish. In the Nettle, which is wind-fertilised, the anthers are provided with a spring which, suddenly opening, scatter the pollen.

In the Polygonaceæ, the species of the genus Rumex are wind-fertilised; occasionally, however, visited by insects.

The species of Polygonum differ considerably from one another in the mode of their fertilisation. Some, as, for instance, *P. aviculare* (Knotweed), have small inconspicuous flowers, and very little, if any, honey. They are consequently but seldom visited by insects. Other species, on the contrary, such as *P. Fagopyrum* (the Buckwheat), and *P. Bistorta*, are much more conspicuous, contain honey, and are fertilised by insects.

These species, however, also differ considerably ; *P. Bistorta* is proterandrous. When the flower opens the stamens are ripe, while the stigmas do not mature till the anthers have shed their pollen, and shrivelled up. *P. Fagopyrum*, on the contrary, is dimorphous ; some plants having short stigmas and long stamens : others, on the contrary, long stigmas and short stamens. In *Polygonum amphibium* the stems, if growing in water, are smooth : while if on dry land they are provided with a certain number of glandular hairs.

The curious arrangement by which cross-fertilisation is secured in Aristolochia, has been already described in the introductory chapter (*antè*, p. 31). Asarum, according to Delpino, is also proterogynous.

Ruppia is an aquatic genus. At the time when the pollen is shed, the female flowers are immature, and the flower-stalk is short and submerged ; when, however, the pollen has all escaped, the female flowers mature, the flower-stalk elongates and assumes a spiral form, so that notwithstanding any slight change of level, the flower rests on the surface of the water. A similar arrangement occurs in Valisneria.

Potamogeton is proterogynous (Delpino—*Ult. Osserv.* Part ii. p. 22).

In the Amentaceæ (oak, beech, willow, poplar. hazel, hornbeam, birch, alder, &c.) the flowers are unisexual, and generally monœcious ; the males are, in some species—as, for instance, in the hazel—visited by insects for the sake of the pollen. As, however, they scarcely ever produce honey, the female flowers offer no attraction to insects, which consequently take no part in the fertilisation.

UPHRYS APIFERA.

CHAPTER VII.

MONOCOTYLEDONS.

IN this class the plumule, or bud, is in germination developed from a sheath-like cavity on one side of the embryo.

Although among the Monocotyledonous orders we do not meet with so many instances of adaptation to insects as is the case in the Dicotyledons; none are more curious or interesting than those afforded by the Orchidáceæ.

ALISMACEÆ.

Alisma Plantago has rather small, pale, rose-coloured flowers, forming a loose pyramidal panicle one to three feet high. The flowers secrete honey from twelve glands, situated on each side of the projecting bases of the stamens. These are six in number, and the pollen-covered side of the anthers is, according to H. Müller, turned outwards. Under these circumstances, insects are more likely to fertilise the flower with pollen obtained from another blossom than with its own.

In Butomus, on the contrary, the flowers are on stalks, and form a large flat umbel. They are proterandrous; while Triglochin, according to Axell, is proterogynous.

HYDROCHARIDEÆ.

This order contains three British genera; Elodea, Hydrocharis, and Stratiotes.

Elodea canadensis (Anacharis Alsinastrum) is a common American weed, which first appeared in our country in 1847, and has since spread with great rapidity. It is dioecious, and it is remarkable that it has not as yet been known to produce male flowers in this country; they are, moreover, rare in America. The female flowers are small, with a long, threadlike, perianth-tube, containing a style which terminates in three stigmas.

Stratiotes aloides is also dioecious. The male flowers are contained several together in a spathe, stalked, and have twelve or more stamens. The female flowers are solitary and sessile. Both sexes secrete honey.

This order is the subject of Mr. Darwin's admirable work, "On the Various Contrivances by which British and Foreign Orchids are fertilised by Insects," from which the following facts are taken. The order contains sixteen British genera, several of them extremely curious and pretty. The species with long nectaries are fertilised by Lepidoptera, those with shorter ones, as a general rule, by bees and flies; *Epipactis latifolia*, it is said, exclusively by wasps, so that, according to Darwin, "if wasps were to become extinct in any district, so would the *Epipactis latifolia*." Other species on the contrary such as *Epipactis viridifolia*, and *Ophrys apifera* (the Bee Orchis) habitually fertilise themselves. It is remarkable that in some Orchids the ovules are not developed until several weeks, or even months, after the pollen tubes have penetrated the stigma. (Hildebrand, *Bot. Zeit.*, 1863 and 1865. Fritz Müller, *Bot. Zeit.*, 1868.)

The flower in this order is very abnormal. There is, except in Cypripedium, only one anther, which is confluent with the style, forming the so-called "column." The anther is divided into two cells, which are often so distinct as to appear like two separate anthers. The pollen in most Orchids coheres in masses, which are supported by a stalk or "caudicle;" the pollen masses with their stalks are called "pollinia." The styles are theoretically three in number; but the stigma of the upper one is modified into a remarkable organ called the "rostellum," and those of the two lower ones are often confluent, so that they appear like one.

Orchis mascula (Fig. 120) is perhaps our commonest species.

Fig. 121 represents the side view of a flower from which all the petals and sepals have been removed, except the lip (*l*) half of which has been cut away, as well as the upper portion of the near side of

FIG. 120.—*Orchis mascula.*

the nectary (*n*). The pollen forms two masses (Figs. 121, 122*a*, and 123), each attached to a tapering stalk, which gives the whole an elongated pear-like form, and is attached to a round sticky disk (Fig. 123*d*), which lies loosely in a cup-shaped envelope or rostellum (*r*). This envelope is at first continuous, but the slightest touch causes it to rupture transversely,

and thus to expose the two viscid balls (*dd*). Now
suppose an insect visiting this flower: it alights on

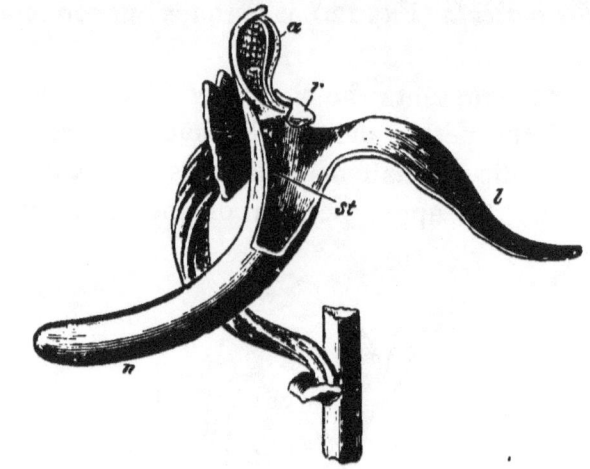

FIG. 121.

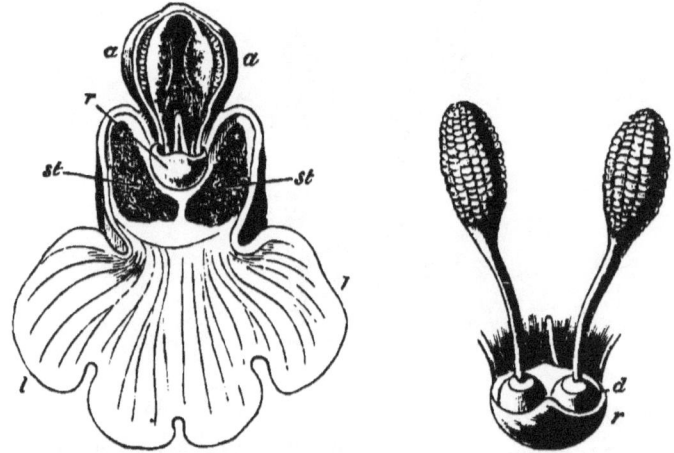

FIG. 122.   FIG. 123.

FIG. 121.—Side view of flower, with all the petals and sepals cut off except the
     lip, of which the near half is cut away, as well as the upper portion of the near
     side of the nectary .
FIG. 122.—Front view of flower, with all sepals and petals removed except the lip.
FIG. 123.—The two pollinia.

the lip (*l*), and pushing its proboscis down the nectary to the honey, it can hardly fail to bring the

base of the proboscis into contact with the two viscid disks, which at once adhere to it, so that when the insect draws back its proboscis, it carries away the two pollen masses. It is easy to imitate this with a piece of grass, and to carry away on it the two pollen masses and their stalks. If, however, the pollinium retained this erect position when the insect came to the next flower, it would simply be pushed into or against its old position. Instead, however, of remaining upright, the pollinia, by the contraction of the minute disk of membrane to which they are attached, gradually turn downwards and forwards, and thus when the insect sucks the next flower, the thick end of the club exactly strikes the stigmatic surfaces (*st st*). The pollinium or pollen-mass consists of packets of pollen grains, fastened together by elastic threads. The stigma, however, is so viscid, that it pulls off some of these packets, and ruptures the threads, without removing the whole pollinium, so that one pollinium can fertilise several flowers.

This description applies in essentials not only to *Orchis mascula*, but also to *O. Morio*, *O. fusca*, *O. maculata*, and *O. latifolia*, as well as to *Aceras anthropophora* (the Man orchis), in all of which the pollinia undergo, after removal from the anther cells, the curious movement of depression, which is necessary in order to place them in the right position to strike the stigmatic surface.

O. pyramidalis differs from the above group in several important points. The two stigmatic surfaces are quite distinct, and the rostellum is brought down, so as to overhang and partly close the entrance to the nectary. The viscid disks which support the pollen masses, are united into a single saddle-shaped body. The lower lip is furnished with two prominent ridges, which serve to guide the proboscis of the insect into the orifice of the nectary. It is of course important that the proboscis should not enter obliquely, for in that case the pollen masses would not occupy exactly the right position.

Following Darwin and other botanists, I have applied to the spur of Orchis the term " nectary." As a matter of fact, however, the flowers of this genus produce no honey; whence Sprengel applied to them the term " Scheinsaftblumen " or " Sham-honey-flowers." Darwin does not, however, think that moths (by which the flowers of this group are principally fertilised) could be so deceived for generation after generation; and as he has observed that the membrane of the interior of the spur is very delicate, and the cellular tissue extremely juicy, he suspected that insects possibly pierce the membrane, and suck the juicy sap lying beneath. His suggestion has been confirmed by H. Müller, and he himself in a subsequent memoir (" Ann. and Mag. of Nat. His.," 1869, p. 143) speaks confidently on the point. Delpino, on the contrary, is confident that the species examined by him (*O. sambucina, O. morio, O. mascula,* and *O. maculata*) do not secrete honey either on or under the epidermis.

The flowers belonging to the genus Ophrys are formed somewhat on the same plan as those of Orchis, but they have no spur, and the rostellum is double. The Bee orchis (*O. apifera*), Fig. 124, however, differs widely from the other allied British species. The two pouch-formed rostellums, the viscid disk, and

the position of the stigma, are nearly the same, but
the stalks of the pollen masses are long, thin, flexible,
and too weak to stand upright. The distance of the
pollen masses from one another, and the shape of
the pollen grains is moreover variable. The anther
cells open soon after the flower expands, and the
pear-shaped pollen masses drop out, so as to hang
directly over the stigma, with which a breath of air
is sufficient to bring them in contact. While therefore in most species of Orchis and Ophrys, self-fertilis-

Fig. 124.—*Ophrys apifera.*

ation appears to be impossible, in the Bee Ophrys,
as R. Brown long ago pointed out (*Trans. Linn. Soc.,*
v. xvi.) it is carefully provided for. Darwin has
examined hundreds of flowers, and has never seen
reason in a single instance to believe that pollen had
been brought from one flower to another; and he

has met with very few cases in which the pollen mass failed to reach its own stigma. He has never seen an insect visit the flowers of this species, and R. Brown suggested that the resemblance of the flower to bees was to deter insects from visiting them. Darwin does not think this probable. He believes also that, though this species habitually fertilises itself, the curious arrangements which it possesses in common with other allied species, are of use in securing an occasional cross, even if only at very long intervals.

Ophrys arachnites is by some botanists (for instance by Bentham) regarded as a mere variety of *O. apifera*, but the stalks of the pollen masses are not much more than half as long, without any diminution of thickness. In proportion, therefore, and in their stiffness, they more nearly resemble the other section of the group. Mr. Moggridge, however, has found at Mentone intermediate forms, not only between *O. arachnites* and *O. apifera*, but also between these, *O. aranifera* and *O. Scolopax*. *O. arachnites* and *O. apifera* do not in England appear liable to pass into one another.

In the Musk orchis (*Herminium monorchis*), the stalks of the pollen masses are short, and the disks large. This species does not produce honey, but has a strong odour, especially at night.

Habenaria chlorantha (the Large Butterfly orchis) has both a sweet scent and honey. It is much frequented by insects. The anther cells are widely separated; the pollinia slope backwards, and are much elongated; the viscid disk is circular, prolonged on its imbedded side into a short, drum-like pedicel.

When exposed to the air this drum contracts on one side, and alters the direction of the pollen mass, thus bringing it (as in *Orchis mascula*) into such a position that it comes in contact with the stigmatic surface of the flower to which it is carried.

Habenaria bifolia (the Lesser Butterfly Orchis) is by Bentham and other high authorities, considered as a mere variety. Yet, as Darwin points out, it differs

FIG. 125.—*Cephalanthera grandiflora*

in many important particulars. The viscid disks are oval ; the viscid matter itself is of somewhat different character ; the drum-like pedicel is rudimentary ; the stalk of the pollen mass is much shorter ; the packets of pollen shorter and whiter ; and the stigmatic surface more distinctly tripartite.

The genus Cephalanthera (Fig. 125, *Cephalanthera*

N

grandiflora) differs from those hitherto described in not possessing a rostellum, and in having the pollen grains single. The flower stands upright, and the labellum is formed of two portions; a base, and a small triangular flap, which at first closes the tube; then turns back, thus forming a small landing place in front of a triangular door, situated half way up the tube; and lastly rises up again and closes the

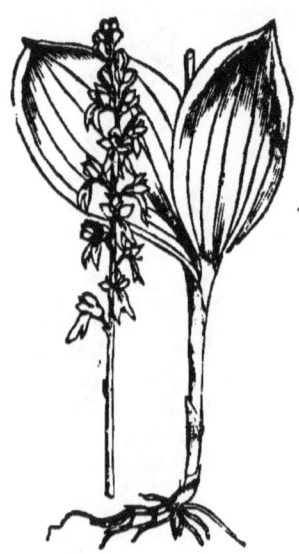

FIG. 126.—*Listera ovata.*

entrance. The pollen mass is situated just above the stigma; and while the flower is in bud, or at any rate before it becomes quite open, the pollen grains which rest on the sharp edge of the stigma, emit a number of tubes which deeply penetrate the stigmatic tissue. These serve partially, but, as Darwin has shown, only partially, to fertilise the flower; he suggests that the principal use of this closing of the flower and emission of the pollen tubes is probably to retain the pollen, which would otherwise fall out of the flower. In this curious manner, however, they are retained in a proper position until the flower is visited by insects, to which they readily adhere; and which are necessary to ensure the perfect fertility of the plant.

Listera ovata (the Twayblade, Fig. 126) has been carefully described by Sprengel, by whom the structure and action of the rostellum was, however, misunderstood, and by Dr. Hooker (*Philosophical Transactions*, 1854), who described the flower with accuracy and minuteness; but the relations of the flower to insects, and consequently the true functions of the various parts, were first perceived by Waechter. The pollen masses lie immediately above the rostellum; the pollen is friable and would not of itself adhere to insects, but this is effected by a very remarkable contrivance (see Hildebrand, p. 53). The moment the summit of the rostellum is touched it expels a large drop of viscid fluid, which glues the pollen to the insect or other body. A very slight touch, even for instance with a human hair, is sufficient to produce this remarkable phenomenon. This species is exclusively visited by ichneumons.

Neottia nidus avis (the Bird's Nest Orchis) agrees in the essential points with Listera, though in the position of the honey, &c., it offers minor differences.

Cypripedium (the Ladies' Slipper, Figs. 127 and 128, *C. longifolium*), the lower lip has the form of a slipper, whence the name. This genus has two fertile anthers, which are rudimentary in other Orchids, while the one which is present in them is represented

by a singular shield-like body. The opening into
the slipper is small, and partly closed by the stigma
and this shield-like body, which lies between the
other two anthers. The result is that the opening into the slipper has a horseshoe-like form, and
that bees or other insects which have once en-
tered the slipper (Figs. 127–8) have some difficulty
in getting out again. While endeavouring to do
so they can hardly fail to come in contact with the

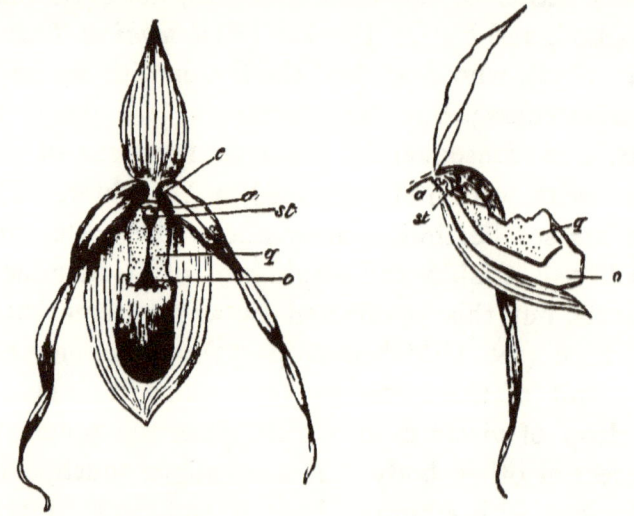

FIG. 127.—Flower of *Cypripedium lon-*   FIG. 128.—Ditto. Seen from the side.
gifolium. Front view.

stigma, which lies under the shield-like representative of the middle anther. As the margins of the
lip are inflected (Figs. 127–8*q*), the easiest exit is
at the two ends of the horseshoe, and by one
or other of these (Fig. 127 *e*) the insect generally
escapes, in doing which, however, it almost inevitably
comes in contact with, and carries off some of the
pollen, from the corresponding anther. The pollen

of this genus is immersed in a viscid fluid, by means of which it adheres firstly to the insect, and secondly to the stigma, while in most Orchids it is the stigma which is viscid.   In a Trinidad species, *Coryanthes macrantha*, according to Dr. Cruger, the basal part of the lip forms a bucket, which secretes a copious fluid which wets the wings of the bees, and by rendering them temporarily incapable of flight, compels them to creep out through the small passages close to the anther and stigma ; thus securing, though by different means, the object which in Cypripedium is effected by the inflected margins of the labellum.   (*Jour. Linn. Soc.*, 1864.)

Such are a few of the remarkable contrivances existing among British Orchids.   I must refer those who wish for more detailed information, to Mr. Darwin's charming work.

Although I have thought it well to confine myself for the most part to illustrations taken from our common wild flowers, I cannot resist mentioning the case of Catasetum, one of the Vandeæ, which as Mr. Darwin says, are "the most remarkable of all Orchids."   In Catasetum, the pollinia and the stigmatic surfaces are in different flowers, hence it is certain that the former must be carried to the latter by the agency of insects.   The pollinia moreover are furnished with a viscid disk, as in Orchis, but the insect has no inducement to approach, and in fact does not touch, the viscid disk.   The flower, however, is endowed with a peculiar sensitiveness, and actually throws the pollinium at the insect.   Mr. Darwin has been so good as to irritate one of these flowers in my presence : the pollinium was thrown nearly three feet,

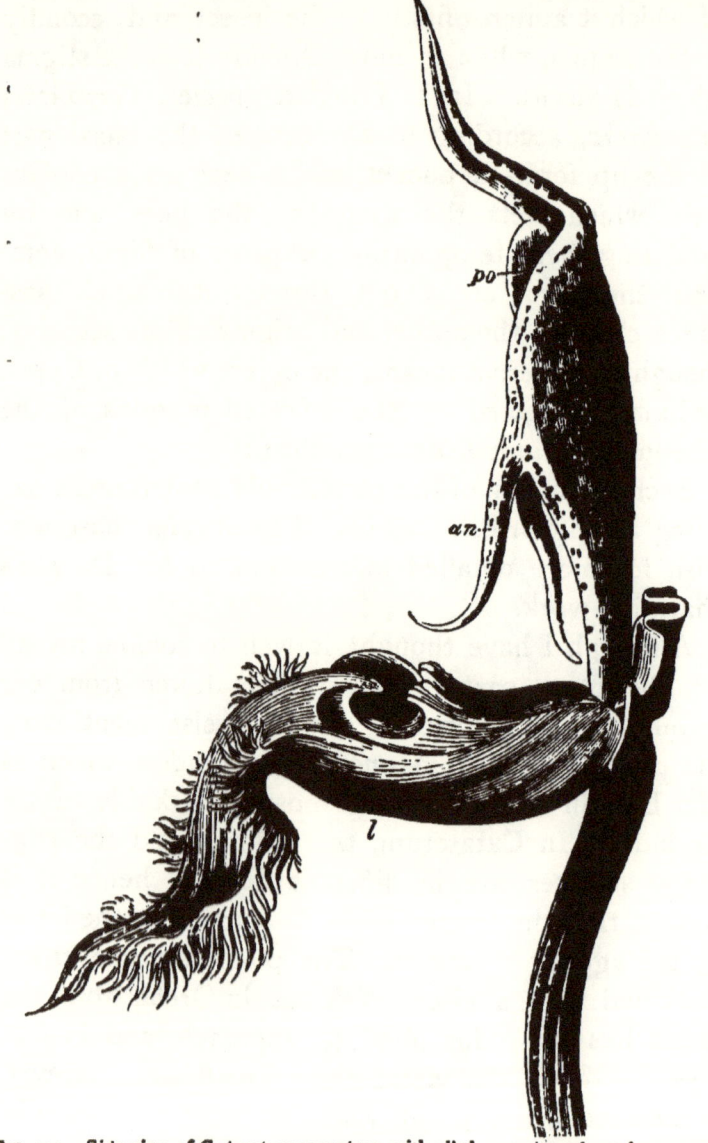

Fig. 129.—Side view of *Catasetum saccatum*, with all the sepals and petals removed
except the labellum.

when it struck and adhered to the pane of a window.

This irritability, however, is confined to certain parts of the flower. Fig. 129 represents a male flower of *Catasetum saccatum*, which is also shown in section in Fig. 130. In this figure it will be seen that the pollinium (*ped*) is curved and in a state of considerable tension, but retained in that position by

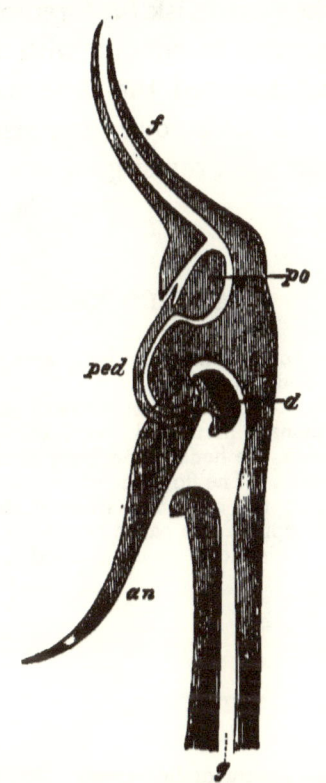

FIG. 130.—Section of ditto, with all the parts a little expanded.

a delicate membrane. Now insects alight as usual on the lip of the flower (*l*), and it will be seen that in front of it are two long processes called antennæ (*a n*). In some species of Catasetum both these antennæ are

highly irritable ; in the present species the right-hand
one is apparently functionless ; but the moment the
insect touches the left-hand one, the excitement is
conveyed along it, the membrane retaining the pollinium is ruptured, and the latter is immediately jerked
out of the flower, by its own elasticity, with considerable force, with the viscid disk (*d*) foremost, and in such
a direction as to come in contact with the head of the
insect which had touched the antenna. On subsequently visiting a female flower the insect brings the
pollen into contact with the stigma.

AMARYLLIDEÆ.

This beautiful order contains three British genera ; Narcissus, Galan-
thus (the Snowdrop), and Leucoium.

The Snowdrop is probably not a true native of this country, but has
long been naturalised in many parts. It is sweet scented, and melliferous ;
as the flower hangs down, the honey is perfectly protected from rain by
the leaves of the perianth. The flower remains open from about ten in
the morning till four in the afternoon, when it closes for the night.
The pistil is white, except at one part a little above the middle where
it is tinged green, a character more marked in the next genus,
Leucoium.

IRIDEÆ.

We have five British genera of this group ; Iris, Gladiolus, Sisyrin-
chium, Trichonema, and Crocus.

Iris pseudacorus L. secretes honey. It is fertilised by humble bees,
and according to Müller, still more frequently by Rhingia. The flowers
are large and showy, the three outer perianth-segments large, spreading
and reflexed, the three inner ones much smaller and erect. The stigmas
are three in number, enlarged, and each with an appendage resembling a
petal, which arches over the corresponding stamen and outer segment of
the perianth. In order to reach the honey, insects have to force their
way between this segment and the over-arching stigmatic leaf.

DIOSCORIDEÆ.

The Yam family contains but one British genus, Tamus ; with one species, *Tamus communis* (Black Bryony). A pretty, straggling creeper, diœcious, with small, yellowish green flowers ; the male in laxer, the female in closer, racemes.

LILIACEÆ.

This order contains seventeen British genera, including the Lily, Onion, Tulip, Colchicum, Asparagus, Solomon's Seal, Fritillaria, Lily of the Valley, Butcher's Broom (Ruscus), &c.

Paris quadrifolia is proterogynous. The perianth is yellowish green, and produces no honey. The structure of this curious flower has not I think been satisfactorily explained. It appears to be one of the species which deludes flies. The dark purple ovary glitters as if it were covered by honey.

The Lily of the Valley (*Convallaria majalis*) is likewise honeyless but is much visited by Hive bees for the pollen.

Allium ursinum is melliferous, and imperfectly proterandrous ; *Llovdia serotina*, on the contrary, is said by Ricca to be very decidedly so.

Hyacinthus orientalis produces no honey, but the fleshy base of the flower is pierced by some insects for the sake of the sap.

The Common Asparagus is a cultivated variety of *A. officinalis*, which grows on maritime sands, or sandy plains, in central and western Asia, and on the south European coasts up to the English Channel. The flowers are melliferous, small, greenish white, on slender stalks two or three together in the axils of the branches. The species is particularly interesting, as an instance of an unisexual flower, which is evidently descended from bisexual ancestors ; since the male flowers contain a rudimentary style, the female flowers rudimentary stamens. In accordance with Sprengel's rule, the male flowers are distinctly larger than the female, being about six mm. long, while the female are only three mm. long.

Colchicum autumnale is proterogynous, though the stigma is still capable of fertilisation when the anthers ripen. Honey is secreted by the base of the stamens.

JUNCACEÆ.

We have two genera belonging to the Juncaceæ (Rushes). Juncus (the Rush), with fourteen species ; and Luzula (the Woodrush) with five. They are wind-fertilised, and, at least as regards some species, are proterogynous.

CYPERACEÆ.

The Cyperaceæ (Sedges) are a very numerous group containing eight British genera. The flowers are minute, greenish or brownish, and wind-fertilised, but are sometimes visited by insects for the sake of the pollen.

GRAMINEÆ.

The order Gramineæ (Grasses) is very extensive, containing more than forty British genera. They are, however, wind-fertilised.

This is the last order which I have to mention. Those who have done me the honour to read so far, will not need to be told that this little book is fragmentary and incomplete. For my own part, I am only too sensible of it. Nevertheless, the fault is not altogether mine. Our knowledge of the subject is as yet in its infancy ; and indeed, my great object has been to bring prominently before my readers how rich a field for observation and experiment is still open to us. Most elementary treatises unfortunately, though perhaps unavoidably, give the impression that our knowledge is far more complete and exact than really is the case. This naturally tends to discourage, rather than to promote, original observations. Few, I believe, of those who are not specially devoted to zoology and botany have any idea how much still remains to be ascertained with reference to even the commonest and most abundant species. In the present case, I have confined myself to the consideration of Flowers in relation to Insects. The interesting adaptations presented by such forms as the grasses, conifers, &c., which are fertilised by the action of the wind, did not therefore come within my subject.

The causes which have led to the different forms of leaves have been, so far as I know, explained in very few cases : those of the shapes and structure of seeds are tolerably obvious in some species, but in the majority they are still entirely unexplained ; and even as regards the blossoms themselves, in spite of the numerous and conscientious labours of so many eminent naturalists, there is no single species as yet thoroughly known to us.

INDEX.

O

THE END.

RICHARD CLAY AND SONS, LIMITED, LONDON AND BUNGAY.

NATURE SERIES.

POPULAR LECTURES AND ADDRESSES ON
VARIOUS SUBJECTS IN PHYSICAL SCIENCE. By Sir WILLIAM
THOMSON, D.C.L., LL.D., F.R.S.E, Fellow of St. Peter's College, Cambridge, and Professor of Natural Philosophy in the University of Glasgow. With
Illustrations. 3 vols. Crown 8vo. Vol. I. CONSTITUTION OF MATTER.
6s.

ON THE ORIGIN AND METAMORPHOSES OF
INSECTS. By Sir JOHN LUBBOCK, Bart., F.R.S., M.P., D.C.L., LL.D.
With numerous Illustrations. Fifth Edition. Crown 8vo. 3s. 6d.

ON BRITISH WILD FLOWERS CONSIDERED IN
RELATION TO INSECTS. By Sir JOHN LUBBOCK, Bart., F.R.S.,
M.P., D.C.L., LL.D. With Illustrations. Fifth Edition. Crown 8vo. 4s. 6d.

FLOWERS, FRUITS, AND LEAVES. By Sir JOHN
LUBBOCK, F.R.S., &c. Second Edition. Crown 8vo. 4s. 6d.

THE TRANSIT OF VENUS. By G. FORBES, M.A., Professor of Natural Philosophy in the Andersonian University, Glasgow. Illustrated. Crown 8vo. 3s. 6d.

THE COMMON FROG. By St. GEORGE MIVART, F.R.S.,
Lecturer in Comparative Anatomy at St. Mary's Hospital. With numerous
Illustrations. Crown 8vo. 3s. 6d.

POLARISATION OF LIGHT. By W. SPOTTISWOODE,
F.R.S. With Illustrations. Fourth Edition. Crown 8vo. 3s. 6d.

THE SCIENCE OF WEIGHING AND MEASURING,
AND THE STANDARDS OF MEASURE AND WEIGHT. By H. W.
CHISHOLM, Warden of the Standards. With numerous Illustrations. Crown
8vo. 4s. 6d.

HOW TO DRAW A STRAIGHT LINE: a Lecture on
Linkages. By A. B. KEMPE. With Illustrations. Crown 8vo. 1s. 6d.

LIGHT: A Series of Simple, Entertaining, and Inexpensive
Experiments in the Phenomena of Light, for the Use of Students of every age.
By A. M. MAYER and C. BARNARD. With numerous Illustrations. Crown
8vo. 2s. 6d.

SOUND: A Series of Simple, Entertaining, and Inexpensive
Experiments in the Phenomena of Sound, for the Use of Students of every age.
By A. M. MAYER, Professor of Physics in the Stevens Institute of Technology,
&c. With numerous Illustrations. Crown 8vo. 3s. 6d.

SEEING AND THINKING. By Professor W. K.
CLIFFORD, F.R.S. With Diagrams. Crown 8vo. 3s. 6d.

DEGENERATION. By Professor E. RAY LANKESTER,
F.R.S. With Illustrations. Crown 8vo. 2s. 6d.

FASHION IN DEFORMITY, as Illustrated in the Customs
of Barbarous and Civilized Races. By Professor FLOWER. With Illustrations.
Crown 8vo. 2s. 6d.

CHARLES DARWIN. Memorial Notices reprinted from
Nature. By THOMAS HENRY HUXLEY, F.R.S.; G. J. ROMANES,
F.R.S.; ARCHIBALD GEIKIE, F.R.S.; and W. T. THISELTON DYER,
F.R.S. With a Portrait engraved by C. H. JEENS. Crown 8vo. 2s. 6d.

ON THE COLOURS OF FLOWERS. As Illustrated in
the British Flora. By GRANT ALLEN. With Illustrations. Crown 8vo.
3s. 6d.

THE SCIENTIFIC EVIDENCES OF ORGANIC
EVOLUTION. By GEORGE J. ROMANES, M.A., LL.D., F.R.S.,
Zoological Secretary of the Linnean Society. 2s. 6d.

A CENTURY OF ELECTRICITY. By T. C. MENDENHALL. Crown 8vo. 4s. 6d.

MACMILLAN AND CO., LONDON.

NATURE SERIES—*Continued.*

THE CHEMISTRY OF THE SECONDARY BATTERIES
OF PLANTÉ AND FAURE. By J. H. GLADSTONE, Ph.D., F.R.S.,
and ALFRED TRIBE, F.Inst.C.E., Lecturer on Chemistry at Dulwich
College. Crown 8vo. 2s. 6d.

ON LIGHT. The Burnett Lectures. By Sir GEORGE GABRIEL
STOKES, M.A., P.R.S., &c., Fellow of Pembroke College, and Lucasian Professor of Mathematics in the University of Cambridge. Three Courses : I. On
the Nature of Light; II. On Light as a Means of Investigation; III. On the
Beneficial Effects of Light. Crown 8vo. 7s. 6d.

CHEMISTRY OF PHOTOGRAPHY. By RAPHAEL
MELDOLA, F.R.S., Professor of Chemistry in the Finsbury Technical College,
City and Guilds of London Institute for the Advancement of Technical Educa-
tion. Crown 8vo. 6s.

MODERN VIEWS OF ELECTRICITY. By OLIVER J.
LODGE, D.Sc., LL.D., F.R.S., Professor of Experimental Physics in University
College, Liverpool. With Illustrations. Crown 8vo. 6s. 6d.

TIMBER AND SOME OF ITS DISEASES. By H.
MARSHALL WARD, F.R.S., Fellow of Christ's College, Cambridge, Professor
of Botany at the Royal Indian Engineering College, Cooper's Hill. With Illustrations. Crown 8vo. 6s.

Others to follow.

MACMILLAN AND CO., LONDON.

EVERY THURSDAY AFTERNOON, Price 6d.
(A Specimen Number, post free, 6½d. stamps.

NATURE:
A Weekly Illustrated Journal of Science.

NATURE contains Original Articles on all subjects coming within the domain of
Science, contributed by the most eminent Scientists, belonging to all parts of the world.

Reviews, setting forth the nature and value of recent Scientific works, are written
for NATURE by men who are acknowledged masters in their particular departments.

The Correspondence columns of NATURE, while forming a medium of Scientific dis-
cussion and of intercommunication among the most distinguished men of Science, have
become the recognized organ for announcing new discoveries and new illustrations of
Scientific principles among observers of Nature all the world over—from Japan to San
Francisco, from New Zealand to Iceland.

The Serial columns of NATURE contain the gist of the most important Papers that
appear in the numerous Scientific Journals which are now publi hed at home and
abroad, in various languages; while longer Abstracts are given of the more valuable
Papers which appear in foreign Journals.

The Principal Scientific Societies and Academies of the world, British and foreign,
have their transactions regularly recorded in NATURE, the Editor being in correspondence, for this purpose, with representatives of Societies in all parts of the world.

Notes from the most trustworthy sources appear each week recording the latest
gossip of the Scientific world at home and abroad.

As questions of Science compass all limits of nationality, and are of universal interest, a periodical devoted to them may fitly appeal to the intelligent classes in all
countries where its language is read. The proprietors of NATURE aim so to conduct
it that it shall have a common claim upon all English-speaking peoples. Its articles
are brief and condensed, and are thus suited to the circumstances of an active and
busy people who have little time to read extended and elaborate treatises.

Subscriptions to "Nature":
Yearly ... ... 28s. | Half-Yearly ... ... 14s. 6d.
Quarterly ... ... 7s. 6d.
To the Colonies, United States, the Continent, and all places within the Postal Union:
Yearly ... ... 30s. 6d. | Half-Yearly ... ... 15s. 6d.
Quarterly ... ... 8s.
P.O.O. to be made payable to MACMILLAN AND Co.
OFFICE: 29 BEDFORD STREET, STRAND